Plumbing

A Guide for the Illinois Apprentice Plumber

(A Step by Step Guide to You in Control and Saving You Money)

Dorothy Laws

Published By **Jackson Denver**

Dorothy Laws

All Rights Reserved

Plumbing: A Guide for the Illinois Apprentice Plumber (A Step by Step Guide to You in Control and Saving You Money)

ISBN 978-1-77485-570-6

No part of this guidebook shall be reproduced in any form without permission in writing from the publisher except in the case of brief quotations embodied in critical articles or reviews.

Legal & Disclaimer

The information contained in this ebook is not designed to replace or take the place of any form of medicine or professional medical advice. The information in this ebook has been provided for educational & entertainment purposes only.

The information contained in this book has been compiled from sources deemed reliable, and it is accurate to the best of the Author's knowledge; however, the Author cannot guarantee its accuracy and validity and cannot be held liable for any errors or omissions. Changes are periodically made to this book. You must consult your doctor or get professional medical advice before using any of the suggested remedies, techniques, or information in this book.

Upon using the information contained in this book, you agree to hold harmless the Author from and against any damages, costs, and expenses, including any legal fees potentially resulting from the application of any of the information provided by this guide. This disclaimer applies to any damages or injury caused by the use and application, whether directly or indirectly, of any advice or information presented, whether for breach of contract, tort, negligence, personal injury, criminal intent, or under any other cause of action.

You agree to accept all risks of using the information presented inside this book. You need to consult a professional medical practitioner in order to ensure you are both able and healthy enough to participate in this program.

TABLE OF CONTENTS

Introduction 1

Chapter 1: Basics In Diy Plumbing 3

Chapter 2: Tools And Aids For Diy Plumbing 9

Chapter 3: Diy Kitchen Repairs 14

Chapter 4: Diy Bathroom Repairs 18

Chapter 5: Diy Toilet Repairs 22

Chapter 6: Diy Repair Of Plumbing Features 27

Chapter 7: Blocked Drains 30

Chapter 8: Common Plumbing Questions 44

Chapter 9: How The Drainage System Works 64

Chapter 10: Pipe Sizes And Different Types Of Drain Pipes 66

Chapter 11: Locating Waste Pipes And Measuring Their Distances 68

Chapter 12: Cleanout Caps 70

Chapter 13: How To Unclog; All Sinks 72

Chapter 14: Understanding Where Your Home's Water Comes From And Goes. .. 96

Chapter 15: Do It Yourself (Diy) Of Plumbing... 166

Conclusion 184

Introduction

This book contains proven steps and strategies on how you can tackle your plumbing problems on your own.

Plumbing chores are usually avoided by homeowners. Minor problems can seem messy and difficult to handle, which is why they immediately call for a plumber to make the necessary fixes.

But DIY plumbing is an easy endeavor for beginners like you, especially when tackling plumbing systems that are not concealed in walls.

You do not have to possess special skills or knowledge in unclogging, repairing, or replacing plumbing fixtures and fittings.

You do not even have to buy expensive plumbing tools, although you may have to buy replacement parts.

Most often, the act of going to the hardware store takes up more time than doing a plumbing project.

Other "difficulties" you may encounter in your DIY plumbing venture include trying to squeeze yourself into constricted spaces, as well as forgetting to shut off the water supply before doing anything.

The most common plumbing repairs are presented in this book, which are just right for a DIY beginner like you.

This book will help you in fixing a leaky faucet, toilet, and joint connection; giving the sink and showerhead added functionality; unclogging shower drains; and replacing your old toilet as well as your shutoff valve.

Thanks again for purchasing this book, I hope you enjoy it!

Chapter 1: Basics In Diy Plumbing

Nature's basic laws are followed when it comes to plumbing, with factors like water, pressure, and gravity being considered.

Understanding how they all work together will not only let you uncover the mystery to plumbing in general, you will also be able to make a number of DIY repairs—and save time and money in the process.

Water In, Water Out

Two subsystems make up the entire plumbing system, and each one is markedly different from the other.

Freshwater is brought in by the supply system, while wastewater is taken out by the drainage system.

Supply System

The Pressure Is On

If you have ever wondered how the incoming water is able to travel wherever it is needed around the house, it is all thanks to pressure.

Enough pressure allows the water entering your home to travel around corners and even upstairs.

It All Boils Down to Hot Water

Your family's cold water needs are instantly met by the water that comes from the main supply.

Your hot water requirements, on the other hand, are not immediately made available to you because another step has to take place.

Cold water has to be carried to the water heater through one pipe, where it gets heated as well as regulated by a thermostat.

Afterwards, the heated water goes through a hot water line, and then travels until it reaches all appliances and fixtures which need hot water in order to function.

Drainage System

It does not really matter if a septic system or a sewer system is used in your home.

Drainage systems basically work in the same manner – gravity wins over pressure.

Unlike the freshwater supply system, wherein pressure is a factor that needs to be taken into consideration, the drainage system depends on how its pipes are positioned.

It's All in the Angles

Waste. All drainage pipes in a drainage system are angled downward. This allows waste matter from your house to be pulled down by gravity.

This downward flow of drainage water is then continued from the sewer line to either a septic tank or a facility for sewage treatment.

Far From Being Simple

Vents. The drainage system sounds simple enough, but there are things that complicate it a bit, such as the clean-outs, traps, and vents.

Those vents are responsible for letting air into your drainpipes, which is why they are placed in the roof in an upward position.

Wastewater cannot properly drain out if not for the air supplied by the vents. It gets stuck in the traps, where it will have to be drained off with a siphon.

Traps. The S-shaped portion of a drainpipe is the trap, and it can be found under all of the sinks in your house. Water flows through from the basin, through the trap, and out of the drainpipe.

Sufficient amount of water is left behind in the trap, though; it serves to prevent gas from the sewer from backing up by forming a seal.

Working Together

These three components (waste, drain, and vent) serve vital roles in the proper functioning of a drainage system, which is why the acronym DWV (drain-waste-vent system) is usually referred to in plumbing.

The DWV system should be complete and functioning to ensure proper flowing out of water along with the waste.

Bridging the Gap

Because they perform different functions, it is important that the drainage and supply subsystems do not overlap.

They still have to have some connection, though, in order to make the whole plumbing system operational.

This is why water fixtures are essential in a plumbing system, because they serve as the bridges through which the two subsystems can work together.

Power of Three

There is another trio involved in the world of plumbing, and they are the sinks, toilets, and tubs. All of them function as fixtures, in the same way that a faucet and washing machine also act as fixtures.

Their main purpose is to make sure that the supply and drainage functions of a plumbing system are separated.

First Lines of Defense

Main shutoff valve. The main shutoff valve should be the first thing on your mind whenever a plumbing emergency occurs. It is important that it is immediately closed, or else your house will get flooded once a pipe explodes.

Individual shutoff valves. You will find that the main shutoff valve does not have to be closed when there is a need to repair some plumbing fixtures.

This is because they already have individual shutoff valves built into them. Still, it helps to ensure that every family member has an idea where the main shutoff valve is located.

Wait for the Go Signal

There are two things you have to consider, though, before you can start working on your DIY plumbing projects:

1) always make sure that the main shutoff or a fixture's water line is not turned on; and

2) know the limitations to plumbing on your own by consulting with the plumbing code official in your area.

Chapter 2: Tools And Aids For Diy Plumbing

Plumbing Tools

You will find that most of the tools commonly used in plumbing are tools that were not actually intended for plumbing.

This is what makes DIY plumbing a more manageable task to handle since the tools to be used are already available in your home. For particular projects, however, you might need some tools that are especially made for plumbing.

Pipe Wrench

This special wrench is something you would want to be available for certain plumbing jobs that require the loosening of pipes. A pipe wrench is perfect for such jobs with its inherent grip and size.

Tightening and loosening of different kinds of plumbing connections is possible with the use of a pipe wrench which is medium-sized and comes with an adjustable feature.

Basin Wrench

Another type of specialized tool that will help you in doing a lot of DIY plumbing in the house is the basin wrench. It has the ability to let you reach those tight spots while working under basins and sinks.

This is because a basin wrench has jaws that are flexible enough to be able to work with nuts of varying sizes.

Plus, these jaws also have the ability to be flipped over to the other side. Because of this second feature of the basin wrench, you will find it easy to turn it without actually having to remove it.

Socket Wrench

The third tool you would need in plumbing is the socket wrench. A number of household repair jobs can be accomplished with the use of a set of socket wrenches, including the installation and removal of shower and tub fixtures.

You will find that a socket wrench is especially handy for plumbing tasks that require you to take out those packing nuts that are hard to remove, particularly if they are already recessed.

In case you would need to remove your old toilet seat and then install a new one, the deep socket wrench is your best bet.

Spud Wrench

The spud wrench is especially helpful for certain plumbing tasks, like taking out an old toilet unit so that you could either fix it or replace it with a new one. The huge pipe that serves as the connection between the bowl and the tank is called a spud.

The spud can usually be found in older toilet systems, wherein hexagonally-shaped slip nuts in extra-large sizes are used to hold it in place. Removing these slip nuts is what the spud wrench is designed to do in plumbing.

There are two types of spud wrench available – the adjustable type that offers more versatility and the non-adjustable type that is characterized by a fixed opening found on each of its ends.

Plumbing Aids

A lot of DIY plumbing repairs can be done by simply using some of your basic tools in the home. There are times, though, when you have to use specialized tools to ensure that your project gets done.

These tools include the plunger, snake, and auger. They are the plumbing aids – handy devices that can help you clear things up without having to call a plumber or use chemical devices.

Plunger

The simple plunger can help you clear out most blockages.

Snake

A snake is used to clear out drain blockages. It usually comes in different lengths in order to tackle different clogging problems.

Most DIY repairs in plumbing can be carried out with a short snake, but a more heavy-duty type might be needed for some drains, which can differ in their depths as well as severity of their blockages.

Auger

The closet auger is actually a more specialized type of snake. Its design is specifically intended to be used in the unclogging of toilets.

Significantly shorter than the basic snake and protected by housing made of metal or plastic, the closet auger is equipped with a crank that you can easily use to carry out your DIY plumbing projects.

Chapter 3: Diy Kitchen Repairs

Taking the Leaks Out of a Faucet

Besides the fact that they never fail to become a source of irritation, leaky faucets should also be handled as soon as possible to avoid the unnecessary expense of having to replace them.

What You Need:

Tools like an Allan wrench, Phillips-head screwdriver, slip-joint pliers, needlenose pliers, and flat-head screwdriver as well as materials like a faucet repair kit, packing grease, and hot/cold water diverter are needed for this project.

What to Do:

1. Reach for the valves beneath the sink so that the water can be turned off. To let the water drain out, see to it that the faucet is turned on. Avoid having small parts fall into the drain by making sure they are plugged.

2. Take out the faucet handle by releasing its setscrew with an Allan wrench. To avoid

forgetting the right order of the removed faucet parts when reassembling them, get them lined up in the order you have taken them out.

3. Take out the chrome bonnet with the use of slip-joint pliers so that the ball and cam assembly can be removed. To prevent scratching the chrome, get some masking tape and use it to wrap around the pliers' teeth.

4. With the use of needlenose pliers, carefully take out the packing pieces and springs. Make sure that any buildup or sediment from the interior of the faucet is cleaned out.

5. Use both hands to hold the spout firmly. Get it loosened by moving it back and forth, after which you can apply more force to take it out from the faucet face. The O-rings and the diverter should then be removed with the spanner and needlenose pliers, respectively.

6. After ridding the faucet face of all traces of sediment or buildup, get the new diverter and O-rings in place. The top O-rings should be

installed first, but make sure to protect the faucet components by coating it with packing grease.

7. Put the spout back by pressing it firmly into place with your hands. The springs and packing pieces can then be slid into place by using the Allan wrench and spanner to lead the way.

8. The faucet ball will now be positioned in its slot. Set in place the cam as well as the cam packing. The chrome bonnet and the handle will then be reinstalled before checking for leaks by turning on the water.

Upgrading the Sink with a Sprayer

Your sink will benefit from the added functionality that a sink sprayer provides.

What You Need:

All you need for this project are two tools and two materials, namely slip joint pliers, basin wrench, sink sprayer, and plumber's putty.

What to Do:

1. Get some plumber's putty (a ¼-inch bead) and apply to the sprayer base's bottom edge. The sprayer base's tailpiece should then be placed in the sink opening.

2. On top of the tailpiece, place a friction washer from below the sink. Use slip joint pliers or a basin wrench for tightening the mounting nut onto the tailpiece after screwing it into place.

Any excess plumber's putty found around the base should be wiped off.

3. The sprayer hose will now be manually screwed onto the hose nipple, which is found on the faucet's bottom part.

Make sure the hose nut gets a good grip after hand-tightening it by following up with a quarter-turn from the slip joint pliers or basin wrench.

Chapter 4: Diy Bathroom Repairs

Getting the Clog Out of a Shower Drain

If your shower drain does not work properly, you only need to follow these DIY tips to get it easily unclogged.

What You Need:

This project calls for the following tools and materials: plunger, wire hanger, flashlight, hand snake, screwdrivers, and a piece of stiff wire, as well as a small trash bag and gloves (latex).

What to Do:

1. Use a screwdriver in removing the strainer so you can check for clogs. Look into the drain opening for hair strands with the help of a flashlight, and then clear them out with a stiff wire.

2. Position the plunger by letting its rubber cup cover up the drain opening. Make sure that the lip of the cup is submerged in water before forcing the hair clog out by moving the plunger handle using up-down motions.

3. In case there are stubborn clogs that the plunger is unable to remove, use a hand snake this time. Drop the cable's end into the drain hole until you can no longer feel the cable moving along. Move the handle continuously by cranking it clockwise until all the clogs have been pulled out.

Breathing New Life into a Showerhead

Installing a hand-held shower adapter allows you to have a better showering experience (since it can more easily get to those hard-to-reach places unlike a fixed showerhead) as well as gives you an opportunity to perform an easy DIY plumbing task.

What You Need:

This project calls only for one plumbing tool (adjustable pliers) and three materials (showerhead adapter kit, masking tape, and Teflon tape).

What to Do:

1. Make sure your work area is all set. To protect your tub's surface from scratches, cover it up using an old towel.

2. With the aid of adjustable pliers, take out the old showerhead. Avoid getting the chrome ruined by using tape to wrap around the pliers' jaws. Wipe away marks left behind by Teflon tape or old plumber's putty on the shower stem's threads with a rag.

3. Place the supplied washer in the diverter's neck. Make sure to follow the directions provided in the kit.

4. Use Teflon tape to prevent water from leaking out of the shower stem's threads before connecting them to the diverter. Ensure that the tape application follows the direction in which the diverter will be turned during attachment.

5. After connecting the shower stem to the diverter and installing the washer in the showerhead's neck, see to it that the showerhead is attached to one of the diverter's necks.

6. Before installing the hose-extender, apply Teflon tape to the diverter's other neck. Then connect it to the hose-extender along with the hand-held shower attachment.

Chapter 5: Diy Toilet Repairs

Doing Away with Toilet Leaks

A leaking toilet should not be a pressing problem if you follow these easy steps to fixing it.

What You Need:

For this DIY project, you will only need slip-joint pliers, a flapper, and a ballcock assembly.

What to Do:

Inspection

1. Perform an inspection of the float and inlet valves by looking into the tank's interior. You will know that the leakage problem is due to the float or inlet valve when the water level goes beyond the overflow tube.

2. Check whether the problem is caused by the inlet valve by flushing the toilet and then raising the float through the rod that holds it as the water level rises. Release the rod once the water stops. This means that there is

nothing wrong with the inlet valve, and the culprit might be the float.

3. Make adjustments to the float's level by working the screw found at ballcock's top portion. If this does not stop the water level from rising above the overflow tube, it is probable that the float is the cause of the toilet leak. A probable reason for this is that the float has developed a hole that prevents it from lying above the water, which in turn prevents it from tripping the inlet valve. In this case, a new rod and float are in order.

4. Check whether the cause of the problem is the ballcock itself. If the water does not stop rising while testing the inlet valve, the ballcock may be broken. In this case, replacing the entire assembly with a new one is the best solution.

Assembly Replacement

1. At the tank's base, locate the supply line that is attached to the ballcock's bottom. With the aid of slip-joint pliers, take out the

ballcock by removing the nut which secures it to the tank.

2. Remove the old assembly by giving it a push from its base, and then put the new one in place by dropping it.

3. See to it that the newly thread-on nut from the tank's base is securely attached with slip-joint pliers. The supply line will then be reattached.

4. Make sure that the new refill tube is properly placed in the tank's interior by clipping it, and that the water is turned on at the shutoff.

Doing Away with an Old Toilet

The task of removing your old toilet as well as putting in a new one can be a great DIY plumbing project.

What You Need:

Replacing an old toilet requires the following tools: needlenose pliers, adjustable wrench, screwdriver, slip-joint pliers, gloves, rag, and sponge. You will need to have these materials on hand: water closet (one- or two-piece), toilet seat, wax seal, and toilet supply tube.

What to Do:

1. Make sure that the supply valve is firmly closed so that the water supply is shut off.

2. Let a significant amount of water from the tank drain away by ensuring that the lever is held down while you flush the toilet.

3. Get lots of water out of the bowl by using a plunger, and then make sure that the excess water found inside the tank is sponged off.

4. Get the caps out from the bottom of the bowl by popping them. The nuts can be easily removed by using an open-ended wrench.

5. Take out the toilet supply tube to disconnect it from the toilet stop.

6. Lift the toilet off the flange after unbolting the tank from the bowl.

7. Use a rag to prevent the entry of sewer gases into your bathroom by stuffing it into the drain opening.

8. Check if there are cracks in the flange by scraping off the old wax while wearing your gloves. It's time to replace the flange if cracks are present.

9. Make sure that the toilet bolts are pointing up when they are placed in the flange. Take the rag out of the toilet drain, and then seal the flange with wax while ensuring that the bolts are held upright by the wax seal.

10. When attaching the tank to the toilet bowl, ensure that the tank bolts are dropped through the proper holes and the bolts are tightened to prevent wobbling.

11. Set up the new water closet by gradually setting it onto the wax seal, and checking that the bolts are lined up with the bowl's mounting holes. Hook up the toilet supply tube to reconnect it to the tank, which should then be refilled to make sure there are no leaks.

12. Use the seat bolts that come with the water closet to get the seat attached.

Chapter 6: Diy Repair Of Plumbing Features

Dealing with a Bad Shutoff Valve

Whether you are getting a new plumbing fixture installed or repairing a new one, you will find that leaks usually occur in the old shutoff valves when they are shut off. That means it is time to replace the shutoff valves as well, especially when turning them becomes difficult.

What You Need:

This project calls for tools such as gloves, adjustable wrench, soldering torch, locking pliers, flame-protecting cloth, and a new shutoff valve (Choose the appropriate type for your water system.).

What to Do:

1. Switch off the main water supply.

2. Turn on the faucet located on top of the shutoff valve as well as a faucet in the basement.

3. Make sure that the supply lines to faucets are disconnected.

4. Take out the old valve as well as the compression nut if your system has compression fittings.

5. Use the soldering torch to remove the old fittings. Protect the flammable surfaces in the cabinet's interior with a flame-protecting cloth.

6. Install the new valve after sanding off any remaining solder.

Making Leak-Proof Connections

A plumbing leak is easily prevented when the fittings for the water supply and waste line are properly connected.

What You Need:

Making sure that your plumbing joints are properly connected will need tools such as a wrench set, slip joint pliers, and an adjustable wrench as well as materials such as Teflon pipe joint compound, Teflon tape, and flexible supply tubes.

What to Do:

Threaded Joints

1. Use three layers of Teflon tape to wrap the male threads in the clockwise direction, making sure that the threaded pipe's end faces you. Increase the number of wraps for loose fittings.

2. On the tape's surface, apply Teflon pipe joint compound to form a thin layer.

3. Before using wrenches to tighten the connection, the threads should be started manually at first.

Compression Joints

1. Use a small amount of Teflon pipe joint compound to make sure the pipe and the ferrule are well-lubricated. This helps you use less pressure in fitting the ferrule into the pipe.

2. Use two wrenches to ensure that the compression fittings are tightened.

3. Check that the pipe is not misaligned with respect to the fitting in order to prevent leaks.

Slip Joints

1. Use a bit of pipe joint compound to get the slip joint and drain tubing lubricated, as well help in letting the washer glide easily and ensuring a more secure seal.

2. Manually get the slip joint started. Continue turning until its threads are properly engaged.

3. Make sure that all joints are hand-tightened before adjusting the trap parts. See to it that the trap parts are properly aligned and positioned at an angle that allows drainage. This will also help prevent leaking.

4. Get the nuts tightened with a half-turn by using slip joint pliers.

Chapter 7: Blocked Drains

First of all you need to know when your drain is blocked. There are a number of clear signs that will make you aware of this. Most notably, a horrible smell will emanate from the drains, this should be pretty easy to spot and it's your first sign that there's a problem. Second keep an eye out for any manholes near your property. A blocked drain will cause these too overflow and this leads to damp patches around your home and garden. Finally if the toilet doesn't flush properly and the waste remains in the bowl you have a serious drain problem. It might seem obvious that these signs point to you having a blockage, however there are some signs that come before these which will help you to stop a problem occurring. It's much better if you can spot a problem before it gets that bad so keep an eye out for these pre-emptive signs.

If you hear a gurgling noise coming from your drains whenever you flush the toilet or empty the sink you have a blockage on the way. Also another sign is if the water levels in your toilet remain lower than normal. Finally a slight smell will be evident from each of the drains on your property. This may start as a slight smell coming from the sink but will progress until it smells very strongly. Look out for these early warning signs so you know whenever you have a blockage building up somewhere in your system.

Spotting these signs will help you get the problem fixed early on so you don't end up with a drainage disaster. Once you've noticed the signs you can try to unblock the drain on your own using the following methods.

Baking soda and drains

There are a couple of ways you can use baking soda to keep your drains clear throughout the year. The first option is to pour a cup of baking soda down the problem drain, and follow it up with a cup of hot vinegar. The mixture will react and fizz, which will hopefully dislodge any dirt and debris and keep your drains from getting clogged up. Follow this mixture with some hot water to wash away anything that remains. If this is the first time in a while you've cleaned your drains then you may need to repeat the process a few times to get the full effect.

Another option open to you is to use 1 cup of baking soda and half a cup of salt. Pour this mixture into the drain and let it sit there overnight if possible. In the morning pour some hot water down the drain on top of the salt and baking soda mixture. This will help to keep the drains clean and prevent a clog from building up.

These same methods can be used to clear out your garbage disposal. You probably won't need to use as much baking soda for this task though. Cut the measurements in half to get approximately the right amount for a smaller drain. Baking soda can even be helpful on its own. Pour some down the drain if you're going away and it will help to prevent smells appearing and dirt building up.

Not only is baking soda much cheaper than normal cleaning chemicals it also works just as effectively. Baking soda has other benefits in that these methods are entirely environmentally friendly. Also using baking soda avoids the need to use toxic chemicals which can have a damaging effect on pipes and even your health. It's always best to use a non-toxic substance if possible, and in this instance baking soda makes it possible.

Plumbing snake

A plumbing snake is a device that helps to unblock your drains when they get clogged with dirt and debris. It consists of a flexible piece of metal that looks much like a spring. This tends to be very thin so that it can fit into any drain that might become clogged. A plumbing snake is designed to be very manoeuvrable so that it can be rotated and moved whilst in the pipe to clear out a blockage.

The plumbing snake is inserted into the drain and pushed along the pipe until it reaches the clog. There are multiple methods of removing the clog. Some plumbing snakes consist of a corkscrew like end which is burrowed into the clog. Once this is secure you can pull out the plumbing snake and this will hopefully bring the clog with it allowing water to flow freely again. Another method is for the plumbing snake to punch holes through the clog making it more prone to breaking up on its own. As the clog gets broken up the flow of water is able to carry some of it down into the sewer system. It may take a few attempts but this should shift the clog with ease and it avoids having to dispose of it yourself.

The advantage of using a plumbing snake is that you can avoid the use of harmful chemicals. These chemicals often wear away at the pipes and can lead to long term damage. A plumbing snake will cause no damage to your system and it can be reused every time a clog occurs. Also having a plumbing snake will save you loads in plumber charges, a plumber dealing with a clog will likely use a plumbing snake anyway. So save yourself some time and money and get one of your own.

Water Heaters

To start with it's a good idea to know what sort of Water Heater you have and the differences between types. This will give you a clear understanding of what is and isn't possible with your particular water heater.

The first and most common water heater is the storage tank water heater, which can be powered by gas, electric or oil. A tank is filled with water and heated so that hot water is available at all times. When hot water leaves the tank, cold water enters and is heated providing a constant supply. Some storage tank heaters are insulated to provide a much higher energy efficiency. They may also come equipped with a more efficient heat exchanger so that less heat is lost from the heat source to the water. These more energy efficient heaters tend to be higher priced but can save as much as 40% on energy costs.

Another common water heater is the tankless water heater also known as an instantaneous water heater. These have no storage tank for the water, instead these systems heat the water on demand only when it's actually needed. This avoids heat loss through the tank and pipes from water just waiting to be used. In their most basic form a tankless water heater consists simply of a heat source such as an electric element which heats the water as it passes by. These are generally only used to supply a specific area with hot water as they don't have the capacity to provide hot water for an entire home.

Finally the integrated space/water heater is becoming more popular with consumers as time goes by. These systems combine both your homes heating needs with its hot water requirements. This means that only one boiler is needed, with one heat source and only one vent. This saves quite a bit of money on installation costs and is less intrusive to the available space in your home. These systems are only really effective when a high efficiency boiler is used. Otherwise the higher running costs quickly outweigh the cheaper installation.

Signs of water heater problems

There are many things that can go wrong with your hot water heater. Knowing the signs can help you to get the problem sorted before it becomes a disaster. To begin with if you notice that your hot water is much colder than usual then you probably have an issue with your drip tube. This is a small tube that lets cold water into the system so that it can be heated. Often a leak will develop with this tube which allows cold water into the hot water supply. This will cause your hot water to be much cooler than it should be.

If you notice your water heater making strange noises this is a sign that sediment is building up inside your water heater tank. When the water heats it creates steam bubbles which get trapped under the layer of sediment. As these bubbles escape they make a popping sound which is probably what you're hearing from your water heater. This isn't a problem in itself but if the sediment is allowed to build up too much it can cause lasting damage to your water tank.

Another common sign is a leaking temperature-pressure relief valve. This valve allows water to escape if the pressure gets too high, which prevents the tank from exploding. For safety reasons it's recommended you test the valve once a year to make sure it's working properly. The problem is that in practice this test will result in the relief valve leaking water from then on. To stop this you just need to purchase a replacement but most people just live with the slight leak.

These signs can either put your mind at rest or alert you to the fact that you have a problem. Knowing is always better than not knowing in this instance as you can deal with the issue before a catastrophic breakdown.

Water heater maintenance

Keeping your water heater well maintained is vital to ensuring it lives a long and productive life. The first thing you can do is to make sure that the heat is never set too high. Your water heater should never be set higher than 120 degrees. Setting the temperature higher than this can put a strain on your water heater which can eventually lead to damage.

Sediment will build up on the water tank due to your home's water supply, this will cause your water heater to run less efficiently. Not only will this cost you more in bills but it will also cause the unit to make more noise when running. Avoid this easily by having your tank drained once a year.

The next piece of maintenance should be to check the pressure and heat release valve that is present on all water heaters. This releases water if the pressure or temperature is deemed too high for the unit to function properly. This is an important safety device which can protect you and your home. Make sure it's functioning correctly by lifting the lever to see if water sprays out, if it does it's working normally.

Next you'll need to check that the anode rod in your water heater for signs of wear. This is normally made from aluminium or magnesium and is placed into the tank of your water heater to prevent corrosion. Lift it out and check how much of the wire core is visible. If there's more than 6 inches showing, or if the rod is coated in calcium you'll be wanting to get a replacement.

Finally check the entire system for any sign of leaks. Keep an eye out for patches of damp where water might be building up. This could be caused by a valve that needs repairing, or an excess of pressure. Either way you need to get any leaks fixed as soon as possible.

Plumbing maintenance

Keeping your plumbing system well maintained can be hugely beneficial to the amount of water you save. The efficiency of your system can be kept really high if it's up to date with maintenance and this will save you money. Furthermore small maintenance issues can turn into severe faults if left alone and this can result in expensive repairs. Follow these maintenance tips to keep your system running properly.

First of all clean the plumbing fixtures in your home. As time goes by the taps and shower heads will become clogged with dirt and debris. Cleaning them can return your water flow to normal which means your system doesn't have to work so hard. Also use some form of cleaning solution to clean out the drains on a regular basis. By doing this frequently you'll stop the build-up of debris and avoid any frustrating system blockages.

Another good idea is to check all of your appliances for leaks. Many appliances that use water will develop small leaks over time. This isn't a severe problem but it's water that you'll have to pay for. Also a small leak can become much worse and stop your appliances from working altogether. By performing a quick check you can get any problems repaired early on.

Finally if you notice a leak somewhere in your system it's a problem that needs to be dealt with by a professional. However not all leaks are visible and there may be a problem that you just haven't detected. A leak in a pipe behind a wall for instance is very hard to spot. Look out for a damp patch forming at the base of walls as this is a good indicator. If you're at all worried you can get a professional plumber to perform a full maintenance check on your system. As a part of this they'll check for any leaks anywhere in the system.

Chapter 8: Common Plumbing Questions

Probably the most common plumbing question of all is whether copper or plastic pipes are better. The simple answer is that in almost every situation copper pipes are better. This is because they are very durable and long lasting. They also don't react to heat as much as plastic pipes do, which can leak in extreme temperatures. Copper pipes can be fitted and tested immediately making them much quicker to install and have up and running. Plastic pipes are prone to being chewed up by mice and often leak if not installed correctly. Copper pipes can clearly be seen to be the superior choice however they do also come with a much higher price tag. If you've got water with a high PH level it may be wise to stick to plastic pipes as copper can get worn down.

Another common question is whether a plumber should be called to repair a leaky faucet. Really it depends on the situation and the severity of the problem. A faucet repair kit can be purchased at most hardware stores and this will give you the tools and instructions you need to complete the repair. However if the tap is spraying, not just leaking, or if you fear that the problem may go beyond the faucet it's best to call a plumber. Faucet repair is easy but any plumbing issue with the pipes will be tricky to attempt without the proper know how.

The last question that often gets asked is how to work out what size plumbing system you'll need for your home. This varies hugely between countries as there are laws governing how many fixtures are allowed and the size requirements of the pipes. So for a major plumbing installation you'll probably want a professional to work out exactly what size is right for you.

Plumbing considerations before renovations

Getting renovation work done on your home can be a pleasant and exciting experience, if you plan it properly. Before you begin renovating you need to think about the plumbing work that will need to be done with it. Leaving this to the last minute can result in huge delays to your project, and you might even damage your existing plumbing system. When doing bathroom renovations one of the first things to think about is your septic tank. You need to find out whether your existing system will be able to handle the changes you're making. If not you'll need to get that upgraded as well. Doing all this in advance can save you a real headache later on when you discover how much work it takes.

The hot water system in your home should be your next consideration when doing bathroom renovations. This is especially important if you're having a new bathroom installed as your existing water heater probably won't be able to handle the extra capacity. Also consider the water pressure available to you. An extra shower will reduce the available water pressure so you might need to get upgraded. The best way to work this out is to ask a plumber to come and do the calculations for you.

There are also considerations to take note of when renovating your kitchen. It might work brilliantly to move the washing machine to the other side of the kitchen but this could mean major plumbing changes. Work out where your pipes are now and try to take this into account when redesigning. It might save you a lot of money if you try to suit your changes around your existing plumbing work. Finally if you're getting new appliances first work out what plumbing they'll need before they can be installed. Get this work done well in advance and you won't be left in a panic.

Septic tank

It's always a good idea to get your septic tank emptied before it becomes necessary but sometimes life gets in the way and it's forgotten. Don't worry, just keep an eye (and a nose) out for these signs and you should be fine. The first thing to be aware of is whether there is a sewage odour coming from anywhere on your property. This should be pretty evident if it's there. Also damp patches in your garden are a sign that the septic tank is leaking or overflowing. Another thing to look out for is noises coming from your drains. When the drains start gurgling it's pretty likely that your septic tank is backed up and needs emptying immediately.

A clog in your septic tank can have many causes but the most common ones are due to solid waste. Any solid waste that gets stuck is likely to cause a clog somewhere along the line. However solid waste may also cause problems after it enters the septic tank. As the levels in the septic tank rise the solid waste rises with it. This can cause a blockage when the solid waste reaches a certain height and blocks of the pipes that lead in and out of the septic system. This is a problem easily solved by getting the systems emptied and cleaned. It is also a good idea to avoid putting too much solid waste into the septic tank. Try not to put oils, fats, or any paper towels into the system as this will often lead to a clog.

If you do find yourself with a clogged septic system there are a few things you can try before calling in the professionals. First you could try using a drain snake to clear out the drain between your house and the septic tank. Clogs often build up here and this can be an easy solution. Next, check inside the tank to make sure none of the pipes are blocked, if so it's time to get it properly cleaned.

Tips for smooth running septic tank

Keeping your septic tank running properly can save you a lot of hassle in the long run. Problems that seem minor can become worse until you're dealing with a very messy disaster. The first thing you can do to avoid this is to make sure that rain water is diverted away from the septic tank drainfield. If the drainfield becomes clogged with too much water it won't be able to properly absorb liquid waste. Also make sure not to overload your septic system. You can help prevent this by checking faucets for any leaks as well as generally conserving water. For instance you could only run the dishwasher when it's completely full, or you could reduce the water used for smaller loads of laundry.

Keep trees and other plant life at a safe distance from your septic tank because when roots start to grow it can severely damage your septic tank. Make sure there are no trees within 100 feet of your septic tank and you should be fine. Next up try to avoid using the toilet as a garbage disposal. Anything like paper towels or nappies will clog up your tank in a very short time. You don't want to have to pay someone to unclog it just because you couldn't be bothered to walk to the bin.

Next up use your garbage disposal as much as possible. The higher end models will be able to grind food waste up much more effectively and this can double the capacity of your septic tank. The smaller particles are easier for the system to digest making the whole thing run more smoothly. Finally it's always necessary to perform routine maintenance on your septic system. It will need to be pumped every 3 to 5 years and you could take that opportunity to see if any repairs are needed before they become too serious.

Garbage disposals

A garbage disposal can make life a lot easier but from time to time it's bound to get clogged up. Follow these tips to get your garbage disposal running smoothly again in no time. First and most importantly turn the unit off, safety comes first. If your garbage disposal is clogged at the blades use a torch to light up the area and see if the problem is visible. If so you can use some tweezers to get rid of the debris and clean out the blades so that they run smoothly. One thing to note is that if your garbage disposal switches itself off unexpectedly, you may have a reset button that you don't know about. Some disposals will have a built in safety device that turns it off whenever something too big gets caught up. Just look for a small button near the disposal that says reset, this should return everything to normal.

A blockage may also occur further down the disposal in the drain pipe. You can try to avoid this by placing things down the disposal slowly rather than just chucking everything in. Anything too hard will likely not get broken up properly and cause a clog further down.

Also banana peels have been reported as one of the most difficult things for a garbage disposal to handle. Don't assume that your garbage disposal can handle anything you throw at it.

The first thing to try when you've got a blocked disposal drain is to run the cold water at full strength. Do this at the same time as switching the disposal on and off. This combination is often enough to clear small clogs. Anything more serious might require you to locate the clog and unscrew the pipe. You can then clean out the pipe with ease and screw it back in again. Make sure to turn off the water and be prepared for a bit of a mess if you have to resort to this method

Toilet clogs

Having a clogged up toilet can be incredibly frustrating. Maybe you don't have a plunger or perhaps the plunger just isn't working. Either way there's another solution that can help you unclog your toilet with much less hassle. To start you'll need a pan of boiling water so get your biggest pan, fill it with water and bring it to the boil. Next you'll need some washing up liquid to squirt into and around the bowl of your toilet. Make sure to use a good amount and cover everything but you don't need to use a whole bottle. The idea is for the washing up liquid to act as a lubricant and loosen up whatever is causing the clog. The thick liquid will sink and make its way to the clogged section of your toilet. Hopefully it will free up the blockage enough so that it can be dispersed.

Now the next stage is to carefully take your pan of boiling water to the toilet in as careful a manner as possible. Slowly pour the water into the toilet in a smooth motion, to avoid splashing boiling water around your home. The force of the water pushing down on the hopefully well lubricated blockage should be enough to clear it.

This may not work straight away so, if you have no success the first time, don't give up. The first attempt may have been enough to loosen your clog but not clear it completely. Repeat the process a few times if necessary and eventually your toilet will be working normally again. This trick can also work on shower and sink drains. Use exactly the same process but use a funnel to ensure that the water goes precisely down the drain. Learning this easy trick should guarantee your toilets and drains never stay blocked for long.

Clean shower drain

The drain in your shower can become a pretty horrible place if left unchecked. This is to be expected because a shower cleans all the dirt and grime of off you and into the shower drain. There are a few things you can do on a regular basis that will stop this grime from accumulating and leave your shower running smoothly. First of all you can help things every time you take a shower. Once your done rinse away any debris that's left around the shower. This includes dirt as well as any bits of shampoo and soap that are left. Also use a shower cleaner to spray the walls and floor of the shower to stop soap scum from accumulating. When this soap scum is cleaned off it can create blockages in the drain, by using shower cleaner you can stop it from building up in the first place. Try to avoid hair getting caught in the drain by picking it up as soon as it falls. Doing this daily maintenance will keep your shower drain from encountering any more severe problems down the line.

There is also some maintenance you can do on a less regular basis but that still needs to be done frequently. Once a week do a thorough clean of your shower drain. This means pulling out the cover, or unscrewing it if necessary, and scrubbing everything down. Remove any dirt that's built up on the cover and scrub the inside of the drain to remove hair and grime. Also make sure the holes in the drain cover are completely clear and there's no sign of a clog.

If a clog does occur take off the drain covers and fill the shower with an inch or two of water. Use a plunger to attempt to unblock the drain. If this doesn't work try pouring baking soda and vinegar into the clogged drain, this should dislodge any debris.

Find a plumbing leak

Have you noticed a leak in your plumbing system that's pouring water into your septic tank, but can't work out where it's coming from? Most commonly an unexplained leak is caused by your toilet by there are other causes which can be explored. The first place to look is the plumbing fixtures in your home. Check all of your taps and showers for a leak including the often forgotten outside taps. A leaking faucet can waste a surprising amount of water.

Next explore the possibility that your toilet is the cause of the leak. To do this close the shut off valve and flush your toilet. Leave it for 5 minutes to properly drain and then check your septic tank. If the water has stopped then you've discovered the problem, otherwise do the same thing with any other toilets in your home.

A good way to check if a leak is coming from within your home is to turn everything off and keep an eye on your water meter. First make sure that nothing is using water anywhere in your home and then make a note of your water meter reading. Come back a couple of hours later and compare the old reading and the new. If you have a leak the reading will have changed and you can be sure you're losing water from somewhere.

If you find there is still a leak but really can't work out where it's coming from, consider whether there's something you've forgotten. One area which is frequently overlooked is the drain line on air conditioning units. Also some homes have a basement drain which will need to be checked for leaks. Other causes could be the drain pipe for a washing machine as well as the pressure relief valve on a water heater. These are usually on a separate draining system but it's possible that any of them could be tied into your home and could be the cause the leak.

Prevent Pipe freezes

Having your pipes freeze up in the cold weather will cause your pipes to break and this can be devastating to your home. The broken pipe itself is a relatively cheap repair the real damage comes from the water. When the pipe thaws the water will escape and cause serious water damage to your walls and floors. Save yourself thousands in repairs by performing some preventative maintenance before the cold hits.

First start off by protecting any pipes that are exposed to the cold. This obviously includes outside pipes but also any pipes in the attic or crawlspaces of your home. These can be exposed to the cold air by any cracks in the walls or roof. Cover all of these pipes with insulation to protect them from the cold. It might also be a good idea to seal the cracks with caulking as this will help prevent the temperature from dropping too low. Another good tip is to leave your cupboard doors open so as to allow the warm air in your home to circulate through the cupboard space. This will help to keep warm any pipes that run through the walls behind your cupboards.

If you're looking for something a bit more effective than insulation you can always consider getting some electric heating tape.

This can be run along the length of the pipe and provides heat whenever the pipe gets too cold.

Unfortunately sometimes pipes do freeze up and you need to know what to do when this happens. First make sure you know where the main shut off valve for your water is. This will be your first point of call if a pipe bursts and should be shut off immediately. To thaw your pipes you can use a hair dryer to gently warm the area closest to the faucet and work your way back to the coldest section.

Choosing a Plumber

There are a few questions you should ask when choosing a plumber to ensure you find a good service. The first question to ask is whether a plumber can provide you with properly completed work. This can be found out by asking for a few different references as well as asking your friends and family for information. This is seemingly quite obvious but you'd be surprised at the number of people that get a terrible plumber off of the internet only to find out their friends know of an excellent local plumber.

The next questions to ask are whether the price for their plumbing services will be fair and whether this plumber has the proper experience to get the job done. These questions have become much easier to answer with the advent of the internet.

Any reputable plumber will have a website which should have examples of their previous work, a price range and testimonials from old clients. This gives you an excellent idea as to whether this plumber will be right for you. It will quickly become clear if there have been any problems with a plumber from the reviews on their website. You should easily be able to find out the variety of services they offer as well so you can choose a plumber with the right skills.

The more experienced a plumber is the more likely he will be to have a much wider range of skills than a new plumber. This is a great advantage because it allows you to have just one plumber to deal with any plumbing problems you encounter. This will allow you to develop a relationship with one particular plumber who you can trust to do the work properly and without too much disruption.

It will certainly be your main goal to find a good plumber for the lowest price you can. However it's worth considering spending a little bit more on getting an experienced plumber. This will ensure you get someone who can complete the job quickly and professionally.

What makes a good plumber

Being a plumber is not an easy job for anyone. You're working with systems ranging in age from brand new to decades old. This means being familiar with a huge variety of components as well as knowing how to fix or replace them. A service plumber is a plumber who fixes the plumbing work in your home. This work tends to be urgent such as having a water leak or the toilet being clogged. A good service plumber must be able to turn up at short notice, work out what the problem is, provide a quote for repairs and then finally fix the problem. The plumbers that are best at this are those who love their job and are prepared to work hard for the client.

Being a service plumber requires a lot of patience and a willingness to continue learning throughout their career. New designs for toilets, water heaters and everything else are constantly being produced. This means you need to learn new skills all the time. For instance many water heaters have an on board computer to regulate temperature. This means plumbers are expected to venture into electronics to be able to fully repair a problem.

A good plumber will also be considerate that they're working in your home. Some plumbers will easily be able to fix the problem but will show no regard for your clean carpets. Finding a plumber that's willing to respect your rules and who ensures your home is left undamaged, is something worth taking the time to do. Finally it's important to find a plumber who you know you can really trust, you're letting this person into your home so you want to know that they're trustworthy. Knowing you have a plumber that will give you a fair quote and do the job properly is one of the most important factors in finding a good plumber.

Chapter 9: How The Drainage System Works

As you've noticed in diagrams A through E, all drainage systems follow a similar pattern That no matter where a particular drain is located, kitchen drain down to the toilet drain, all of the drain pipes slant in a downward position and lead to the larger 4", 6", or 8" sewer pipe. Which in tum runs into the city sewer usually at the street. This is very important to remember.

Try to envision a tree, first there is the large trunk, then there are the smaller branches that lead upward. If you select any of the upper smaller branches and follow them from their top downward, you will soon see that no matter where they are located they all lead into the larger trunk of the tree. Drain systems work the same way.

With that in mind, I'll go on further to explain the drain systems purpose. First and main purpose is to dispose of any unused water through a city sewer system, that leads into a city refinery, where the water is processed, cleansed and redistributed for further use.

This is the newer method, the older method was to run the drainage into septic tanks, which when full, would need to be pumped out and again be reused. Though the new method is very effective, many people still use the septic tank, (or cesspool), for irrigation purposes. When a septic tank is full, the whole house will not drain and therefore must be emptied or pumped out. .

Here is how the drain system works. As the water goes through the opening of the drain, it then goes through a P-trap,(better known as a goose' neck, U-bend or elbow). The purpose of the P-trap means just like it sounds, to trap the sewage gasses from entering back into your home. These gasses can be very dangerous if inhaled. Alter the water passes through the P-trap, it then flows in a downward direction through the length of the waste pipe and then into the larger 4", 6", or 8" main sewer pipe. And the water then flows, again and still in the downward flow, and empties into the city sewer. Also, if a drain is not properly vented the water will not drain.

Chapter 10: Pipe Sizes And Different Types Of Drain Pipes

Until recently most drain systems were installed with galvanized steel or black iron pipe. This however has created many problems. As the steel ages, it begins to rust on the inside, leaving less space for the water to pass through, and in most cases, forming a blockage especially if you are using a garbage disposal. The food has a tendency to catch right where the rust forms.

It was not until recently that contracting plumbers began to install the black plastic or ABS waste pipe. The life of this pipe is much longer than the galvanized steel or black iron pipe.

The main sewer pipe which runs from the foundation of the building to the city sewer is usually made of red clay and its sizes range from 4" to 8", depending on the number of drains installed. Most residential buildings are 4" pipe. These pipe sizes are fairly simple to acknowledge. All homes, apartments, mobile homes and commercial buildings, beginning fiom where the water drains into the pipe until it reaches the city sewer goes from smaller to larger. Example; a kitchen sink drain will begin with a 2" pipe until it reaches the main sewer pipe which is 4" in diameter.

Chapter 11: Locating Waste Pipes And Measuring Their Distances

To get a good idea of where drain pipes lead and how they work, look under a raised home or building. One where the first floor drain or water pipes are exposed. If you are an apartment owner or renter, and the apartment is over a garage or carport, look at the ceiling of your parking area. Most likely the drain pipe will be exposed. While looking at the exposed drain pipes of the raised house or apartment, with the drain pipes in view, find out exactly where the above drains are. For instance, follow the kitchen sink drain underneath the house or apartment where you will find the 2" waste pipe and follow it until it joins the 4" main sewer pipe that is located directly under the toilet. Do this with all drains. Tub, bath sink, washer, etc. By doing this you will have a better insight of their locations and distances.

Distances vary on different buildings. Another way to judge the distance and locations is by looking at the roof vents. Vent pipes are usually the same size as the drain pipes. You can judge the distance of any utility to the 4" main sewer vent by pacing your feet. This will give you a fairly accurate distance. (Always add on an extra 10 to 15 feet, they do not always lead straight to the 4" pipe).

Always remember that the toilet is the beginning of the 4" sewer pipe. Not all residents have only one toilet, so if you are not sure about which toilet is the closest to the city sewer, you can call the city and they will tell you where to locate the 4" main sewer pipe.

Chapter 12: Cleanout Caps

A cleanout cap is an opening made available for cleaning out drain pipes. By all rights every separate drain should have one. Although I have encountered many drains without one. When it comes to cleaning a drain, a cleanout cap takes a big part as an easy access. If one cannot be found don't worry, there is always another way.

In the following pages I have included where the cleanouts can be found. Not all cleanout caps are the same. The most used cleanout is a round male threaded screw in plug with a square head on it for using a pipe wrench. Another kind of cleanout cap is what is called a no-hub cleanout. This particular cap is the kind that fits over an open drain pipe with a rubber cap and has two large hose clamps tightly secured around the rubber cap and the pipe itself. To remove it, all you need is a small screwdriver or a small wrench. Just unscrew-one of the clamps and lift the cap off the pipe.

Another kind of cleanout cap, which is found in most commercial buildings, is a round screw in cap with male threads, and instead of a square head, it has a long slot, countersunk into the surface of the cap like a large screwdriver slot. These caps take a special tool to remove them, but by using a hammer and chisel it can be removed by hitting the outside edge counter clock wise.

Cleanout caps are the same size as the waste pipe it is being used for. If a cleanout cap is so tight that a pipe wrench cannot loosen it, take a piece of pipe and slip it over the end of the pipe wrench to extend the handle. This will give you more leverage. If it is still too tight, use your foot to push down on the pipe extension. There is more strength in your legs than in your arms.

Chapter 13: How To Unclog; All Sinks

When a sink drain is clogged or backed up with water, there are certain things which must be taken into consideration. The first thing is to determine precisely where the obstruction or clog is. And then the easiest way to get to and remove it. To find out where an obstruction in a drain pipe is, the best way I have found is to fill the sink up with water. If it is already filled, fine. Next locate the cleanout. The cleanouts for most kitchen sinks are usually found extending out of the wall underneath the sink either inside or outside of the residence, and always below the drain itself. Look around you will probably find it somewhere in line with the vent pipe. Or if the house is on a raised foundation, the cleanout could be found extended out of the screen of the vented square or just inside of the vent opening, under the house attached to the 2" waste pipe coming from the sink. If the cleanout for the sink is located inside the house, be sure to spread out some sort of covering, canvas, plastic or newspaper, covering at least a six foot area and keep a bucket handy. Drain cleaning can get very messy. The next step is to loosen the cleanout plug with a pipe wrench just enough to see if water seeps through. If water does to begin to seep through, put a bucket tightly against the wall beneath the cleanout and unscrew the plug a little letting the water drain into the bucket. When the bucket is almost full, re-tighten the cleanout plug, then empty the bucket. Do this as many times as necessary until all the water is drained out of the pipe. Remember that a clogged drain could hold several buckets of water and if a cleanout plug is suddenly removed, all the water up to the level of the cleanout will come rushing out. So if the cleanout is found on the inside of the residence, remove it slowly and cautiously.

If after you have removed the cleanout plug and there is water in the waste pipe, then the clog is somewhere in the waste pipe between the cleanout and the 4" main sewer pipe.

But if after you have removed the cleanout plug and there is no water in the waste pipe and the sink is still full of water, then the clog is between the sink drain and the cleanout.

Another way to tell if the clog is before or after the vent and the ground waste pipe is by blowing air into the sink vent pipe fiom the roof. To do this, first locate the vent pipe belonging to the sink that is clogged, then place a rag around the edge of the vent, not covering the opening. Then take a deep breath of air and exhale heartily into the vent. The purpose of this is to find out if there is any water in the ground waste pipe after where the nipple and the "Y" joins the 2" waste pipe. If by applying air pressure into the vent pipe and the air rushes through meaning there is no water in the ground waste pipe, then the clog is before that vent pipe, either in the sink drain itself, the p-trap, the nipple, or the angle joining the vent pipe.

But if when applying air pressure into the vent pipe and the pressure returns to you and you hear water in the pipe below, then the clog is in the waste pipe between where the nipple connects to the vent pipe and the 4" main sewer pipe. I Still another way to determine if the clog is between where the nipple connects to the vent pipe and the 4" main sewer pipe is to drop a small stone into the vent pipe and listening to hear if the stone hits water. Providing the vent is not offset. If the vent pipe is straight up and down, you will have no trouble hearing whether or not there is water in the waste pipe below.

Not every residence has the same type of sink, and with each type there are certain ways to locate the clog. I have listed each one separately.

A) Beginning with the double sink without a garbage disposal. If only one side of the sink is clogged, the clog is between the sink drain itself and the "Y" leading into the vent pipe. B) Double sink with garbage disposal. The double sink with a garbage disposal can be unclogged the same as the double sink without a garbage disposal,.providing the clog is not in the garbage disposal itself. This requires a special turning tool and possibly maintenance of the garbage disposal itself.

C) All single sinks. If no water is found in the waste pipe after removing the cleanout or by applying air pressure into the roof vent pipe, then the clog is between the drain itself and the "Y" going into the vent pipe.

D) Single sinks with a garbage disposal. The same as the double sink with a garbage disposal with the exception of only one waste pipe going into the wall under the sink.

E) Back to back sinks. Double or single. Back to back sinks are found mostly in apartments or bathrooms that are one behind the other, I.e., one apartment water utility directly behind another or water utility connected to the same supply pipe.

If no water is found in the waste pipe see A & B above. If water is found in the waste pipe past the "Y", then the clog is between the "Y" and the 4" main sewer pipe. The easiest way to get to and unclog an obstruction.

Depending on where each clog is, there are many ways to reach them. I have listed the different accesses, according to what each situation offers. There are also different methods to try for the actual drain cleaning process, and I have included them also.

After locating where the clog is, we will go on with the actual process of clearing or removing the clog from the waste pipe, finding the easiest and best access, what to do with the clog and the right tools for the job. Please keep in mind that while performing any drain cleaning, as with any job, the idea is to do the job properly in order to keep the drain open. A poorly done job will only result in having to do the job over again. But as I have said before, "experience is always the best teacher".

Any clog located in the 1-3/4" or 2" waste pipe between the cleanout (or vent) and the 4" main sewer pipe requires the right length of cable (snake) for clearing the clog out of the pipe. The cable must be as long or longer than the pipe itself. If your only access is through the roof vent, then you must also account for the extra feet. I.e., 20 extra feet for every one story house.

If you do not own a cable, hand or machine type, one could be purchased at a nearby hardware store, plumbing store or rented from an equipment rental store. There are two types of cables, hand operated or machine operated. The drain cleaning machine cable is fairly expensive if one were to be purchased so I would recommend renting one if necessary. (See page II for cable sizes for each size pipe). Renting a drain machine only costs a fraction of what a serviceman would charge to come to your residence. Have the renter show you how to properly operate the machine.

Once the clog has been located and you have acquired a cable, you are ready to open the clogged drain pipe. First determine the easiest opening closest to the obstruction, be it a cleanout, through the vent or by removing the pipe underneath the sink. Try to get as close to the vent as possible. Then be sure to bend the end of the cable before putting it into the drain pipe. The bend should be no more than 1 to 1-1/2". There are many 90 degree bends in a drain pipe and bending the end of the cable allows the cable to make the 90 degree tums. Bending the cable can be done by taking a large pair of pliers and grasping the end of the cable to make the bend. A 45 degree bend should be more than enough. Any more could break the cable. Make sure the bend stays in the cable. There is no need to use blades or cutters in such a small pipe, It would only create problems or could break off in the pipe and possibly get stuck. A bend in the cable is good enough. Wherever you choose to insert the cable, make sure it is going down the waste pipe and not up. You can usually tell by listening into the wall. Another way to be sure is to insert the cable with the bend at the end of the cable angling down. Insert the cable and begin rotating. Take your time and go through the whole waste pipe to the 4" main. You will come across some rough spots and this could be either the obstruction or a 90 degree bend in the pipe. Go through these slowly and after passing these, retrieve the cable a couple of feet and go through it again. If the cable does not seem to be going in apply some pressure, not too much, but as it is turning push in. (be sure to wear gloves). If too much pressure is applied to the cable, the cable could get stuck in the pipe and being stuck it would most likely kink or twist with the motor running right where you are holding it. Once a cable becomes kinked or twisted it can become dangerous. It could wrap around some part of your body, hand or hair and cause severe injury. Therefore always use caution when operating these machines. Never operate any drain machine under a house. This doubles the danger.

Continue to run the cable the entire length of the waste pipe. You will feel when the cable reaches the 4" main sewer pipe when the cable becomes very loose or easy to push through because it has gone into a bigger pipe. After running the cable the entire length of the waste pipe, retrieve the cable. If you had trouble getting the cable through the first time, due to a heavily clogged pipe, repeat the process again to make a bigger opening in the pipe. Then run hot water in the drain. The purpose of this is to flush any loose particles into the city sewer and to check the drain to see if it is unclogged or not. Make sure that all openings are closed first. If the blockage has not been cleared, all the water will come out the removed pipes or the cleanout opening. If after running the water and the drain is open, it is unclogged. If it is not un-clogged and the water backs up again, re-check for the obstruction and repeat the process until the drain is cleared. Unclogging obstructions located between the drain of the sink and the vent pipe behind the wall.

Double sink - First determine if both sides are clogged or just one side. If both sides are clogged after removing the cleanout, then the obstruction is located in the vent pipe below the "Y" or in the "Y" itself. To unclog this you would have to remove the p-trap on either side of the sink drain and then remove the nipple to gain a close view of the "Y". Insert your cable or a long flexible object into the "Y" making sure it goes down into the vent pipe and not up or a crossed to the other side of the sink drain. Push down until you feel the obstruction then rotate the cable or long object until it has scraped the pipe and removed the clog. Always be sure the whole section of pipe is opened fiom the "Y" to the cleanout. Then re-assemble the removed parts and check your drain. If only one side of a double sink is clogged, the obstruction is either in the p-trap, under the "Y" or in the "Y" of the side that is clogged. To clear the obstruction, first remove the p-trap. Always check the p-trap for softness or rot. Never push hard on any old p-trap. This could result in purchasing a new one. Check the p-trap, nipple, and the "Y" for the obstruction and when locating the clog, remove it, reassemble and check the drain.

If it is a single sink, and there is no water in the drain pipe past the cleanout or the vent pipe past the wall, then the obstruction must be in the p-trap or the nipple. If only one side of the double sink is clogged, follow the instructions above. Unclogging a washing machine drain; This book only covers the drain itself. Anything from the hose that empties the water from the washing machine is not considered the drain. When emptying a washing machine and the water that drains out, backs up in the standpipe,(drain), or even if it starts to back up, then there is a clog in the drain or one forming. The easiest way to locate the obstruction is to first fill the washer with water. Preferably without any clothing in the machine. Then turn the dial directly to the spin cycle, while keeping your hand on the dial ready to turn to the closest off position the second you hear water filling up in the standpipe, (drain).

While water is draining from the machine into the drain, listen carefully into the drain to see how long it takes the water to back up. If it takes say, a half tub of water before it backs up, then the obstruction is located in the 2" waste pipe leading into the 4" main sewer pipe. To unclog this it would require the use of a drain machine with no larger than a 1/2 " cable the length or longer than the waste pipe. To insert the cable into the waste pipe, you must either remove the cleanout cap, go in through the roof vent or remove the stand pipe and p-trap. Always make sure the cable goes down the drain and not up the vent. Also remember to put a small bend at the end of the cable. Slowly run the cable the whole length of the drain pipe, retrieve and repeat for better clearance in the pipe. Then replaced all removed parts or cleanout plug and test the drain with the washing machine full of water and on the spin cycle.

When emptying the water from a washing machine and the water immediately backs up the stand pipe or washer drain, then the obstruction is in the standpipe, the p- trap, or the nipple between the p-trap and the vent pipe behind the wall.

If the p-trap is exposed, or can be reached for removal, take it apart and clean out the inside with a screwdriver or a long object being careful not to puncture or damage the p-trap itself. Also clean the nipple adjoining the vent pipe and the "Y" that joins the vent pipe. And at the same time checking for the clog. Some standpipes are behind the wall and this would require renting a drain machine with a W' to 3/8" cable. Always be sure there is a bend in the front end of the cable before using it.

Then insert the cable directly into the top of the stand pipe opening and carefully run the cable back and forth through the p-trap, nipple and "Y", and then down the 2" pipe itself. When finished test the drain with the tub full of water and on the spin cycle. This also flushes all the loose debris. of a drain it would be wise to clean the whole waste pipe. Most washer drains adjoin with the kitchen sink drain,

Unclogging a tub drain;

Out of all the drains I have encountered, I find the bathtub the most complicated to unclog. But with a little patience and determination it can be accomplished. First I will explain the different lay outs. There are residents with one tub, there are residents with two tubs, one downstairs and one upstairs directly above one another, and there are residences with two tubs back to back and possibly another located elsewhere in the house either upstairs or downstairs. These back to back tubs are found mostly in apartments. (See diagrams).

The cause of most tub clogs is usually hair. Which catches mostly between the drain itself and the connecting vent pipe.

All bath tubs should have an opening to gain access to the p-trap. I will go on further describing the single and back to back bathtubs, how to locate an obstruction, and the different methods of removing such clogs.

Single tubs;

The first thing to do is to determine where the clog or obstruction is located. There are certain steps to follow as such, check the drain screen cover and the drain opening itself for any blockage. If this is not the problem go on the roof and apply air pressure into the vent pipe belonging to the tub drain. If the air pressure returns, guarantying that water is in the 2" waste pipe leading to the 4" waste pipe, then to clear or remove the obstruction you would need a drain cleaning machine with a 1/2" cable the length of the entire drain pipe including the length of the vent pipe on the roof. If there is no water in the drain pipe while applying air to the vent pipe, then the obstruction is located between the tub itself and the nipple adjoining the vent pipe. Most bathtubs have cleanout access for cleaning out tub drains or there p-traps. After locating the cleanout access, remove and using a flashlight, study the p-trap and connecting pipes. If the p-trap is in a position to remove, do so slowly being careful not to bust the pipes. It is advisable to empty as much of the water as possible out of the tub before opening any cleanouts or pipes. Also remember to place a floor covering underneath and keep a bucket handy. Alter removing the p-trap, check it for the clog, then check the nipple adjoining the vent pipe and check the pipe leading to the tub drain itself. Also check the tub drain. Any hair blockage will usually begin forming there. After finding the obstruction and removing it you can then reassemble the removed parts and check the tub drain with water. This procedure can also be done if the house or building is on a raised foundation or crawl space.

If there is no way to get to the p-trap and remove it, you could either try a hand snake or motorized drain cleaning machine with a 1/4" cable. Put a' 1" bend in the end of the cable. Insert the end of the cable downward through the tub overflow inside the bathtub, pushing down until you reach the p-trap. Begin operating the machine and feed the cable slowly through the p-trap, nipple and into the 2" waste pipe. When you have reached or passed the obstruction, the water will properly flow out of the tub. It is advisable to run the machine through the p-trap a couple of times to guarantee a longer lasting clearing. When retrieving the cable, remove it slowly. Nine out of ten times the obstruction will come out on the end of the cable.

Back to back bathtubs, which are usually found in apartments or two bath houses, one bathroom behind another, are similar to single tubs with the exception that they connect to the same 2" vent and waste pipe. When having drain problems with these back to back tubs, first check to see if both tubs are not draining. If they are both clogged, renting a motorized drain cleaning machine would probably be required. Insert the1/2" cable into the roof vent pipe and run the entire length of the waste pipe until you can hear the water draining out of the waste pipe. If only one side of the back to back tub is clogged, see section on single tubs.

Unclogging a shower drain;

When a shower drain is clogged, the problem again is usually hair. Rent a 1/4" or 1/2" cable drain cleaning machine. Remove the screen over the shower drain and insert the cable into the shower drain itself through the p-trap the nipple and then the entire length of the drain pipe to the 4" main. You will know when you hit the obstruction when the water, starts to drain down. If it is a back to back shower and both sides are clogged, rent a 1/2" cable machine and run it through the roof vent.

Floor Drain;

First determine where the obstruction is by locating the cleanout and removing it or by applying air pressure into the roof vent. If there is water in the cleanout or the roof vent, rent a 1/2" cable drain cleaning machine and run the cable the cleanout if there is one or through the roof vent. If there is no water in the cleanout or the vent, then the obstruction is between the floor drain itself and the cleanout or vent pipe. To unclog this, rent a 1/4" or 3/8" cable drain cleaning machine and run this down through the floor drain itself until the water drains out.

Urinals; They are the same as the shower or floor drains. Please see urinals in the illustrated section.

Unclogging the 4" main sewer drain;

You will easily know when the 4" main sewer pipe is clogged when every drain in your house is backed up with water. For instance when you run the washing machine and the toilet begins to bubble and the water from the washing machine backs up into the tub, Immediately shut off all appliances that use water and don't run any water at all. It could cause a flood.

The main sewer line is the large 4" pipe that all of the other drain pipes connect to. So if the 4" main sewer pipe is clogged all of the other drains that are connected to it will be backed up with water also. The first thing to determine is where the obstruction is located. This is done by locating the cleanout for the larger 4" pipe. This could be found in a number of places, such as, in the bathroom behind the toilet, on the floor, outside the house on the ground or near the toilet. They are usually 2 feet from the building or above the ground extending off the wall of the building right behind the wall of the toilet. If your house is on a raised foundation it could be extending out of the 4" pipe under the house and directly under the bath room toilet. After finding the cleanout remove the cap. If water is standing in the 4" pipe or is at least visible, the obstruction is further down the drain pipe in the direction of the street.

There are many reasons why the 4" pipe could be clogged, food, rust, grease, wadded up paper or even tree roots. To clear this pipe would require renting a larger drain cleaning machine with a 5/8" or l" cable. Measure the distance from the cleanout to the street so you will know how long of a cable you will need or how far down the pipe to run it. On this particular job, the cable should have the blades (three cutting blades on the end of the cable), Just make sure that the blades are sharp enough to cut through any roots or any other obstruction that may be clogging the pipe.

Insert the cable into the cleanout always making sure that it is aiming to the direction of the street. Begin operating the machine. Let the cable feed itself into the pipe, always being prepared to shut ofi' the machine in an instant. Perform this operation slowly and cautiously. When your cable reaches the obstruction the cable will begin to tighten up. This means that it is cutting. Let it go, but if it gets too tight, pull back on the cable until it spins freely again and let it feed itself. I repeat again that this must be done slowly and cautiously until the cable reaches the city sewer, pushing the obstruction out of the 4" pipe and into the city sewer.

If you cannot find a cleanout for the 4" main sewer pipe there are two other options, either remove the toilet from off the floor and run the cable into that floor opening, or run the cable through the roof vent above the toilet. Just make sure that the cable is long enough to reach the street. Again always be careful.

If you have separate bathrooms and one is flushing ok but the other one is backing up into the tub, then you have a clog in the 4" main pipe somewhere between the two toilets. So what you want to do is to get behind the obstruction and push it out towards the street. If there is a cleanout for that toilet, then run the cable through it. But if there is no cleanout then either go in through the roof vent or remove the toilet that is backing up and run the cable down through that opening.

 Toilets;

You will know if there is an obstruction in the toilet itself when you flush it and the water backs up right away. And there are different ways to unclog these. By using a plunger, or by renting an auger (a hollow tube with an inner cable with a hand crank on the end). Insert the auger directly into the toilet and run the length of the cable through the built in p-trap and down through the bottom hole of the toilet. In most cases this should do the trick. If this doesn't and you have attempted several times without progress, the toilet will have to be removed. This can be done in this order, First plunge or empty out as much water as possible, then shut off the water supply to the toilet, which is usually directly under the toilet tank. Next unscrew the water supply line at the tank bottom and remove it. Then unbolt the base bolts of the toilet at the floor, being careful not to crack or break the toilet and then lift the toilet off the floor. If the 4" pipe in the floor has water in it then the obstruction is not in the toilet but in the main pipe itself and you will have to run your larger cable through it. But if there is no water in the 4" pipe the clog is in the toilet itself. And the best way to unclog this is to lean the toilet over on its side and run your auger in through the bottom of the toilet back up through the top and pushing the object out from where it originally went in, because the opening at the bottom of the toilet is smaller than the opening at the top. Sometimes just rolling the toilet upside down will free the object letting it fall out back through the top opening.

Before you remove the toilet it sometimes helps to insert a small mirror into the seat of the toilet down into the top hole and looking up you may be able to see the object that is blocking the hole.

After removing the obstruction and before replacing the toilet, you should always install a new wax ring. This serves as a gasket from the floor pipe and the base of the toilet so water does not leak out. Reassemble the toilet in the same fashion as you removed it. Turn the water back on and try flushing it. Always be careful handling the toilet as they are easily broken and are very expensive.

Chapter 14: Understanding Where Your Home's Water Comes From And Goes.

Every home with plumbing has a supply of cold water coming into the building and a main drain to handle sewage leaving the building. The incoming cold water supply feeds both the cold water faucets and taps, as well as the water heater.

Hot and cold water supply lines extend to most fixtures in your house, with a series of smaller drain pipes leading from sinks, toilets, showers and tubs to the larger main drain. In general, problems with the water supply side of residential plumbing are usually about leaks of some kind. Drain issues are usually about blockages. Kitchen sinks and toilets are the most likely location for blockages, but all drain pipes can become blocked.

TECH TIP: Handling a Major Plumbing Leak

Although major plumbing leaks don't happen often, when they do, they can cause serious damage in a very short time. That is why every homeowner needs to know where the main water supply shutoff valve is and how to use it. Find the main water supply pipe that enters your house, typically located on a basement wall. If your house is in a subdivision, water supply lines usually enter the wall closest to the road. The first valve present on this pipe is the main shutoff. Rotate the handle clockwise until it stops to shut off the flow of water. The valve handle might be difficult to turn, but clockwise rotation is the way most main valves shuts off.

Turning off your water supply

For most major leaks you will need a professional plumber to fix the issue, but being able to turn off the water supply beforehand can help to avoid additional damage. It is recommended that everyone in your household know how to shut off the main water supply valve in case of a sudden leak.

BATHROOM PLUMBING — What To Know

With a high concentration of plumbing fixtures, bathrooms are often the source of plumbing issues in your home. Most household toilets, sinks, showers and bath tubs are used frequently so occasional issues are not uncommon. Understanding how these plumbing fixtures work is the first step in understanding what is causing the problem.

Toilets

Whenever the flush handle of a toilet is pushed down, it opens a flush valve in the bottom of the water tank. This allows water to rush down into the toilet bowl, carrying waste into a large drain pipe hidden in the floor. As the water level in the tank drops during a flush, it opens a fill valve to refill the tank to a preset water level, ready for the next flush. The water that remains in a toilet bowl after flushing seals out sewer gases and prevents them from entering your home.

Components of a toilet includes:

- the bowl
- seat
- lid

- water tank
- flush valve
- fill valve
- flush handle

Potential trouble spots for a toilet include the water supply connection to the water tank, the gasket that seals the toilet to the floor drain, the gasket that seals the tank to the bowl, and the flush valve and fill valve inside the tank.

How to choose a new toilet

Not all new toilets flush as well as they should, and that's why you should consult independent testing to choose an effective model. An organization called MaP publishes performance results from hundreds of toilets available around the world. Click here for free recommendations on finding a new toilet that flushes cleanly.

Toilet Problems and Solutions:

Clogged or slow flushing toilet

If you can't clear a blocked toilet with a plunger, a plumber may be required to use a snake to clear the blockage. Mineral buildup inside the bowl of an older toilet can also cause lazy flushing of an otherwise unblocked toilet. Toilet replacement by a licensed plumber is the best option for older toilets that regularly flush poorly.

Constantly running toilet

A leaking flush valve at the bottom of the toilet tank is one common cause of water that keeps running continually into a toilet tank and bowl. A defective or poorly-adjusted fill valve can also allow water to continue filling the tank without stopping. There's limited risk of water damage involved in these repairs, so some homeowners tackle it themselves. A licensed plumber is also an option.

Water supply leak

Even a small water leak from the pipe or hose that connects to the toilet tank is serious because it can develop into a large and damaging leak. The water supply will need to be shut off, the connection repaired, then the water turned back on. Most homeowners require a plumber for this task.

Tank-to-toilet leak

If a leak appears where the toilet tank connects to the back of the bowl, the tank-to-bowl gasket needs to be replaced. The tank will need to be drained and the tank removed from the toilet bowl to install this gasket. Most people hire a plumber for this work.

Base-of-toilet leak

If water appears where the toilet sits on the floor, a defective toilet ring is probably the cause. Sometimes the water only appears immediately after a flush. The toilet tank will need to be drained, the toilet bowl unbolted from the floor, lifted, a new ring installed, then the toilet and tank replaced. This repair is as complicated as toilet replacement, so most people hire a plumber for the work.

Overflowing toilet

Most toilets have a valve on the pipe connected to the toilet tank and shutting off this valve will stop a toilet from overflowing. If your toilet doesn't have a shutoff valve, shut off the main water supply valve for your house. Don't turn this valve back on until the toilet blockage is cleared and the toilet bowl is emptying properly.

Bathroom Sinks and Faucets

These are made differently than sinks and faucets in other parts of your house and they have unique repair issues.

Elements of a bathroom sink and faucet installation include:

- faucet
- sink bowl
- mechanical drain stopper
- water supply lines
- drain pipe

Potential trouble spots for bathroom sinks and faucets include defective cartridge or washer, clogged or slow drain, leaking drain, and malfunctioning drain stopper

How to choose a new Bathroom Sink and Faucet

These are usually replaced as part of a bathroom renovation or vanity cabinet replacement. Bathroom sinks can be a molded part of the vanity cabinet top, they can be under-mounted to the bottom face of the countertop or they can be mounted on top of the vanity top resting in a hole cut on site during installation. You'll need to get your plumber involved early on in your bathroom renovations to determine if pipe and drain locations need to be changed.

Bathroom Sink Problem and Solutions:

Dripping Faucet

Cartridge or washer needs to be replaced. Handy homeowners can do this work themselves, but older faucets often have corroded parts that can break during removal.

Stop and call a plumber if you can't get things to come apart. Some faucet manufacturers offer free replacement cartridges or even free replacement faucets. Call customer service before you buy any repair parts.

Clogged or slow drain:

This is a common bathroom sink problem because soap products and loose hair sometimes build up within bathroom drains. If your sink has a mechanical drain stopper, check to see if the blockage is caused by hair buildup on the horizontal pivot rod a couple of inches below the drain opening. This common problem can be remedied by lifting out the drain stopper and removing hair with your fingers or needlenose pliers.

Leaking Drain Pipe

Water pooling on the bottom of a bathroom vanity cabinet or the floor underneath an open sink is usually caused by leaks where the drain pipe meets the sink, a leaking drain trap plug, or flaws in drain pipe joints. Occasionally under-sink leaks are also caused by a loose or broken drain stopper mechanism or by failed caulking joints that allow splashed water from the countertop to leak in around top-mount sinks. Repairing drain leaks can be more challenging than replacing a faucet cartridge so most homeowners call for professional help.

Malfunctioning drain stopper

This mechanism allows the drain to be opened and closed with a knob at the top of the faucet. It's typically a reliable part of a bathroom sink installation, but problems can still occur. A loose retaining nut securing the horizontal pivot rod can prevent the drain stopper mechanism from working or it can allow waste water to leak under the sink. Tightening the ring around the horizontal pivot rod with your fingers can sometimes eliminate leaks. Replacement drain stopper mechanisms can be purchased separately for DIY repairs or you can call a plumber.

Deteriorated sink caulking

Many bathroom sinks are molded parts of the vanity countertop so there are no caulked joints to leak. But if your sink sits on top of a countertop, and caulking is missing or deteriorated, splashed water can pool and leak around the sink and under it.

Sink smells like rotten eggs

Assuming your bathroom drain was installed correctly, odours like this are caused by microbial infection of the drain and possibly the sink overflow passage. A treatment of 3% hydrogen peroxide from a drug store could solve this problem. Pour 500 ml (2 cups) of hydrogen peroxide down the drain at night. If the odour persists the next morning, remove the drain stopper and stuff a rag down the drain to block the overflow passage with the top of the rag remaining above the drain opening. Fill the overflow passage with hydrogen peroxide and let it sit for two hours before removing the rag.

Showers

This heavily used part of most bathrooms includes a waterproof enclosure with hot and cold running water delivered by a faucet that's typically enclosed behind the shower wall.

Elements of a shower and faucet include:

- faucet
- enclosure
- drain
- shower door or curtain

Potential trouble spots for a shower include the faucet valve, drain and the enclosure itself.

Shower Problems and Solutions:

Dripping shower head

Most showers have a single valve that controls hot and cold operation and a worn valve that allows dripping is the most common shower issue. A handy homeowner can replace the valve after shutting off the water supply, but most people call in a licensed plumber for the work. As with sink faucets, some manufacturers offer a lifetime warranty on shower valves. Call customer service to see if they'll send you a replacement valve for free, even if you're hiring a plumber to install it.

Clogged or slow drain

Soap and hair buildups are the most common cause of drain problems in a shower. Drain cleaner may work in mild cases, but a plumber is usually required.

Leaking shower enclosure

This problem is not uncommon. Shower enclosures can leak where the floor meets the walls, or where a door joins to the shower opening. Some showers leak through tile grout on inadequately constructed showers. Although caulking may be able to stop leaks around doors, most leaky shower enclosures need to be rebuilt.

Smelly drain

This is caused by the same microbial infection that makes bathroom sink drains smell. Pour 500 ml (2 cups) of 3% hydrogen peroxide down the shower drain at night before bed. If the problem is microbial, the odour should be gone in the morning.

Mold growth

Increased ventilation is the cure for this problem. Run exhaust fans for at least 20 minutes after each shower and leave shower doors and curtains open between uses. Mold-resistant paint is effective at discouraging growths on walls and ceilings.

Bathtubs

Most tubs offer the opportunity to take either a bath or shower and feature a combination faucet that includes taps, a tub spout and a shower head. Access to the hidden valve assembly is sometimes found through the wall on the other side of the tub.

Elements of a bathtub installation include:

- faucet and shower head

- bathtub
- door or shower curtain
- drain

Potential trouble spots in a bathtub installation include a leaky water supply valve, clogged drain or defective enclosure or caulking.

Bathtub Problems and Solutions:

Dripping faucet or shower head

Bathtub faucets and shower heads are part of the same plumbing fixture and operate from the same single or double valves. Replacement is an option for handy homeowners if shut off valves are present, but most people call a plumber for this work.

Clogged or slow drain

Soap and hair can cause a restriction in a bathtub drain just as it can in a bathroom sink or shower. Drain cleaner is worth a try, but typically a plumber is required to clear stubborn bath tub drain blockages.

Smelly Drain

As with sink and shower drains, a bathtub drain can develop the odour of rotten eggs if it becomes infected with microbes. Pour 500 ml (2 cups) of 3% hydrogen peroxide from a drugstore down the drain at bed time and the odour should be gone by morning.

Damaged caulking

All inset bathtubs rely on caulking to seal the joint between the top of the tub and the surrounding walls. The most difficult part of re-caulking an existing tub is removing the old caulking. Solvents are available to make this easier. Ensure the area to be caulked is clean and dry before following the instructions to apply the new bathroom caulking. Some homeowners hire a handyman to remove & replace the caulking. The neatest way to apply caulking yourself involves laying down strips of masking tape 3mm or 4mm away from the centre line of the joint. Lay down a bead of caulking, smooth it with a rubber gloved finger dipped in a mixture of dishwashing liquid and water, then peel the masking tape off. The edges of the caulked joint will be perfectly straight and neat.

Mold growth

Increasing ventilation is the way to stop the growth of mold in your bathtub. Run the bathroom exhaust fan for at least 20 minutes after a bath or shower and leave the shower curtain or doors open after use. Mold-resistant silicone caulking is made especially for use around tubs and showers.

KITCHEN PLUMBING — What to know

The kitchen is the centre of most homes, and plumbing has a lot to do with the reason why. Preparing and serving food and cleaning up afterwards depends largely on plumbing. Kitchens also have the greatest number of water-related technologies at work. In addition to faucets and drains, you might also have a dishwasher, a fridge with water dispenser, and maybe even laundry equipment in the kitchen.

Sink & Faucet

Elements of a kitchen sink and faucet installation include:

- double or single sink bowl
- drain pipe and debris screen
- faucet
- spray hose with head*
- liquid soap dispenser*
- hot beverage water dispenser*
- reverse osmosis drinking water tap

*Optional features, may or may not be present in your kitchen

Potential trouble spots for a kitchen sink and faucet include dripping faucet (most common), leaks where the sink meets the counter top, leaking drain pipes, leaking connection between the water supply pipes and the faucet, a clogged drain, or a malfunctioning garbage disposal or hot water dispenser.

TECH TIP: Washer or Cartridge? What's the Difference?

Most faucets and taps made until the 1970s used a small disk of rubber called a "washer" to stop the flow of water when the hot or cold tap handle was shut off. The faucet mechanism closes down and squeezes against the washer, sealing off the space for water to flow through. Eventually the washer gets old, brittle and cracked, allowing a small flow of water even when the handle is tightened down fully. Inexpensive replacement washers are still available in any hardware store to stop dripping, even for very old faucets. Most modern faucets don't have washers but use a replaceable valve cartridge instead. When these leak, the only practical solution is to replace the cartridge. Both washer and cartridge replacement is something that a handy homeowner could complete themselves.

Kitchen Sink & Faucet Problems and Solutions:

Dripping faucet

Most modern kitchen faucets use a single valve cartridge to control the flow of hot and cold water instead of the washers used in older faucets. Replacing the cartridge stops dripping of either hot or cold water supply. Some leading manufacturers offer a lifetime warranty on residential plumbing faucets, including free replacement cartridges. Contact the manufacturer to see if they will ship a new cartridge at no charge. Internet images are a good way to determine what model of faucet you have.

Clogged or slow drain

This is a common problem with kitchen sinks because of the food debris and grease that sometimes goes down the drain. Drain cleaner liquid can sometimes speed up a slow drain, but many times using a sink plunger or drain snake is necessary. Small plungers made for use with sinks are easy to use, but don't use a plunger if you've already poured drain cleaner down first. Corrosive liquid could splash upwards at you. If you are using a plunger, simply place the mouth over the open drain, then push the plunger handle up and down with quick movements. If this doesn't work, a drain snake is a coil of stiff metal meant to poke through blockages and open up the drain pipe. The need to open the drain trap to get into the drain with a snake means that most people call a plumber if drain cleaner or a sink plunger doesn't work.

Leaked water on the bottom of the sink cabinet

The most common cause of this problem is a leaking drain pipe joint. Even if you don't feel comfortable doing a drain pipe repair, take a look under the sink with good light to determine exactly where the leak is coming from. You'll be able to hire a plumber with more confidence if you know where the problem is coming from.

In-cabinet leaks can also be caused by a faulty water supply connection to the faucet. This is a much more serious leak than just a dripping tap because it can lead to a massive and damaging house flood if the leak gets worse. Modern faucets usually have flexible hoses that connect water supply pipes to faucets, and the fittings that make these connections can be tightened by hand. Shut off the water supply valve feeding the line, then tighten the fitting by hand clockwise. Turn the water back ON again, looking carefully for leaks. If tightening the fitting by hand doesn't solve the problem, call a plumber. Leave the water turned off until a repair is made.

Splashed water can also leak down into the sink cabinet if the caulking or seal around the sink is defective. Removing and replacing sink caulking can be DIY work, or you can hire a handyman. Plumbers won't usually do a house call to simply repair faulty caulking, though most will apply caulking if they're installing a new sink.

Malfunctioning hot water dispenser

This small, separate tap dispense steaming hot water for instant use making coffee, tea, hot chocolate or instant soup. Not many homes have this kind of fixture, but if repair or replacement is required it's best done by a professional since the work involves plumbing and electrical tasks.

TECH TIP: Gurgling Drains Solutions

The lack of an air vent is the leading cause of a sink drain that makes gurgling sounds as the water goes down. Adding an air admittance valve to the drain pipe may solves the problem. Not every jurisdiction allows air admittance valves, but they can work well and are relatively easy to install. If you are experiencing a gurgling drain, call a licensed plumber to see if an air admittance valve is an option.

Dishwasher

All dishwashers have a connection to the drain (usually under the kitchen sink), but most only have an inlet for a hot water supply. Since dishwashers don't have cold or warm cycles, a hot water feed is all that's needed. This hot water feed usually taps into the supply pipe near the kitchen faucet. The only exception to this are high-end dishwashers with internal water heating capabilities. These might have a cold-only connection or they might heat the already-hot water to a higher temperature.

Elements of a dishwasher installation include:

- the dishwasher appliance itself

- a flexible hose carrying pressurized water (usually hot only) to the inlet of the machine

- a flexible drain line leading from the dishwasher to the kitchen drain pipe.

Dishwasher Problems and Solutions:

Dishwasher not cycling properly

There are different issues that can cause a dishwasher to fail to run well, some of which include plumbing-related issues. The most common cause of dishwasher problems is a clogged drain screen with too much trapped food on it. This prevents waste water from being pumped out of the machine. All dishwashers have at least one filter, and cleaning them is a regular household maintenance task for homeowners. Some dishwashers have filters that clean themselves, or at least try to. If your filters are manual, clean them after every few loads.

Sometimes a dishwasher won't even start if the filter is too dirty.

Dishwasher not cleaning or draining properly

Besides a clogged filter, a malfunctioning drain pump or drain water sensor could be the cause. While plumbers typically install new dishwashers and deal with water supply and drain issues, appliance repair technicians are the people who solve mechanical issues inside the dishwasher.

Dishwasher Door leaks water

A pool of water on the floor in front of the door can be caused by overloading the machine so the door doesn't close properly, but more commonly it indicates a malfunctioning door seal or drain pump. You can check the seal yourself by looking for cracked, folded or broken sections that allow water to leak, or call an appliance repair technician.

TECH TIP: Dirty Dishes Solution

If your dishwasher fails to get your dishes, glasses and cutlery as clean is you'd like, and it appears to be running well in other ways, try running your kitchen faucet until the water is hot before each dish washer load. These days dishwashers use so little water that they might not actually get to the completely hot water from the hot water supply pipe. Running the kitchen hot water tap until the water gets hot can improve dishwashing results by eliminating the residual cold.

Fridge with Water Dispenser

More and more refrigerators have a water supply connection for dispensing cold drinking water or making ice. No drain connection is present with fridges.

Elements of a fridge with a water connection include:

- A small, flexible copper or plastic tube that delivers water to the fridge at the back.

- A replaceable filter that removes unwanted taste from the water

- Potential trouble spots for a refrigerator with a water dispenser include water leaks at the back of the fridge, low or no water pressure at the dispensing nozzle, and failure to produce ice.

TECH TIP: Dating Refrigerator Filters

Depending on how much drinking water you take from your fridge, the filter may need to be changed every six months to a year. Most fridges have a warning light to alert you when filter needs changing, but you might not get around to this right away. Be sure to check the manufactures recommended replacement cycle and mark the installation date on the filter label before installation so you know at a glance how long it's been in place.

Fridge with Water Dispenser Problems and Solutions:

Leaking water connection

A loose or broken water supply connection at the fridge or a broken water supply pipe is the usual cause of leaks from a water-connected fridge. A shut-off valve can be found where the thin fridge supply line connects to a cold water pipe in the basement or under the kitchen sink. Learn where this valve is located so you can shut it off quickly in the event of a leak.

Slow or no water delivery from dispenser

This could be caused by a blocked or clogged supply, or from a water supply pressure that's too low. A plumber can help you with water supply problems to your fridge, but an appliance repair technician will be needed to fix problems inside the fridge.

Failure to make ice in the freezer

This could be caused by a blocked or clogged water supply, but it might also be that the ice maker or the valve controlling water flow into the ice maker is bad. If your fridge delivers water at good pressure from the dispenser, but does not make ice, make sure the ice maker is turned on, and that the metal ice level arm is tilted downwards to the run position. If none of these things get the freezer to make ice, call an appliance repair technician for help. It's not unusual for ice maker mechanisms to fail. They can often be replaced as a single unit at reasonable prices.

BASEMENT PLUMBING - What to know

Most unfinished basements have minimal plumbing, while finished basements can have almost as many plumbing fixtures as above-ground spaces in your home. Plumbing elements that are unique to basements may include:

- main water supply line
- floor drain
- laundry sinks and washing machine
- water supply to furnace humidifier
- outdoor faucet water supply lines
- drain line from heat recovery ventilator (HRV)

Potential water supply and main drain issues in the basement includes leaking taps, faucets and fixtures, main drain sewage backup, leaking washing machine connections, seasonal outdoor tap maintenance.

Basement Plumbing Problems and Solutions:

Leaking basement taps, faucets and fixtures

Repair issues like these are the same as for any tap or faucet. A dripping tap usual requires a replacement washer or valve cartridge, just as with any other tap or faucet in your home.

Main water supply line and valve

Every home serviced by a municipal water supply will have a pipe entering the house (typically a 3/4" diameter line in the basement), a shutoff valve and a water meter. It's not enough to just find the valve, but actually try closing it. Most main valves are round knobs that close with clockwise rotation. Try closing the valve to make sure it hasn't gotten too stiff to move.

If your home is serviced by a private water system, you are responsible for maintaining the pump, pipes, switches, valves, gauge and pressure tank. Has your water stopped flowing? Check to make sure the fuse or breaker supplying the pump isn't tripped. Replace or reset as needed. All private water systems should also have a pressure gauge near the pressure tank. If the gauge shows pressure between 30 pound-force per square inch (psi) and 60 psi, your system is delivering water. Call a plumber familiar with private water systems for help.

Leaking washing machine connections

Washing machines are the only appliance that connects to water supply pipes with two short hoses that have female garden hose fittings on both ends. One hose threads onto the hot water supply pipe valve and hot water inlet port on the washing machine, and the other hose connects to the cold supply valve and the cold water inlet port on the washer. All four of these connections require washers and sufficient tightness for leak-free operation. If any of these connections are dripping water, use a pair of pliers to tighten the fittings clockwise no more than 1/8 turn past finger tight. This can be a bit tricky, so it's recommended that only handy homeowners try this themselves. If the leak persists, shut off the water supply valve, unscrew the garden hose fitting completely, replace the washer, then do the connection back up again, snugged up with pliers.

Insufficient water supply to the washing machine

Assuming the rest of our house has sufficient water pressure, a lack of water at the washing machine could be caused by two things. Most modern washing machines shut themselves down and display a warning code if insufficient water pressure is present. Check to make sure both hot and cold water supply valves are turned all the way ON. Most washing machines also have a screen filter where the water supply hoses connect. Shut off the water supply to the washer, remove the garden hose fittings from the machine, then look inside the port for debris on the inlet screens. Use tweezers and a toothpick to remove the debris. If you have hard water, mineral build up could also be partially blocking the filter screen. Apply vinegar to the filter screen with a spray bottle, let it sit for 30 minutes to dissolve the build up, then use an old tooth brush to remove the minerals. Re-establish the hose connections, turn on the water and test washing machine operation. If your washing machine still displays a code for insufficient water, remove the hoses from the washer again, put the ends in a bucket, then turn the water ON to check flow rate and pressure. If both hot and cold flow seems normal, call an appliance repair technician to look at your washer. If water flow and pressure really are low, have a plumber investigate the cause.

Basement floor drain sewage backup

If you see water bubbling up from the main floor drain in your basement, it's a serious issue. This could be caused by a blockage in the main drain pipe leaving your house, or the entire municipal system could be saturated. This happens most often after heavy rains and will cause sewage to back up in some or all homes in the neighbourhood. If the problem is isolated to your home only, call a plumber and avoid flushing toilets or putting any water down the drain until the problem is resolved. If the problem is caused by a saturation of the municipal system, there's nothing much anyone can do until the wet conditions subside. The ultimate protection against sewage backup is to have a backwater valve installed on your main drain pipe in the basement floor. This one-way valve allows waste water to leave your home, but not to come back inside.

Outdoor tap maintenance and management

There are two kinds of outdoor taps: one operates in all four seasons, even during cold winter weather. The other outdoor tap is a three-season installation that only operates when temperatures are above freezing. The year-round tap has a valve inside the home with a long handle mechanism that extends to a knob outdoors. When you turn the valve ON or OFF, you're turning a shaft that controls the valve that's inside where it's above freezing. You know you have a year-round outdoor tap if you can see an air vent extension off the top of the valve, just behind the valve handle. In order to work properly during winter, the water in this year-round valve must be able to drain out completely. Nothing can be connected to it. This is necessary to allow water inside the valve to drain out immediately after use. If you leave, say, a garden hose connected to a year-round tap, trapped water will freeze inside the valve and break it.

Every three-season outdoor tap needs to be drained in the fall, then re-activated in the spring. Start by finding and shutting off the shut-off valve controlling the tap. This is typically in the basement or some other heated part of the home. Next, go outside and open the outdoor tap all the way. Go back indoors, then find and open the small drain cap on the body of the shutoff valve, if your valve has a drain. Opening this drain cap allows the water in the pipe to come out, emptying that portion of the pipe that extends outdoors. Have a small container ready to catch the water as it comes out of the open drain cap. turn the shutoff valve back ON.

Leaking or blocked HRV drain line

A heat recovery ventilator (HRV) is a permanently installed ventilation appliance that's now common in new home construction. During operation in winter, HRVs generate internal condensation that must drain away. A plastic tube extending from the bottom of the HRV is the usual approach. The amount of condensed water generated is not large, but it can leak from a faulty connection between the tubing and HRV. Also, the tubing can sometimes get plugged, preventing proper draining from the HRV.

OUTDOOR PLUMBING FIXTURES - What to know

Most homes have a least one outdoor tap delivering cold water, but additional outdoor plumbing fixtures are becoming more common. These include hot and cold running water outdoors, and built-in lawn sprinkler systems.

Elements of outdoor plumbing include:

- hose tap (could be standard or frost-free type)
- yard irrigation system
- rooftop plumbing drain vent
- septic system
- greywater pit

Potential issues with outdoor plumbing include malfunctioning outdoor tap or yard irrigation system, a blocked plumbing vent that affects all drains in the house, or a failing septic system or greywater pit.

Outdoor Plumbing Problems and Solutions:

Dripping or inoperative outdoor hose tap

Most outdoor taps use washers instead of cartridges, but either of these may need to be changed to stop a dripping outdoor tap. Find and close the shutoff valve to the tap before working on it. If you can't find the shutoff valve, shut off all the water to the house with the main valve. If outdoor taps become too stiff to open or close, try operating the valve repeatedly to loosen the action. If this doesn't help, replacement of the valve by a licensed plumber will be required.

Outdoor roof vent causing slow or noisy drains throughout the house

This can be caused by a blockage of the main vent pipe, usually where it exits the roof. This happens most often in regions that get very cold winters. Frost builds up inside the vent pipe as it extends above the roof, and over long cold spells the frost can build up and block the rooftop vent from admitting air. The same thing can be caused by rust building up on the inside of older-style steel vent pipes. If all or most of the drains in your house are slow or emit gurgling sounds as water goes down, you may have a blocked rooftop vent.

Yard irrigation system doesn't work

Built-in lawn and garden irrigation systems include permanently buried plastic hose connected to small sprinklers, watering nozzles or trickle irrigation lines across your yard. These systems deliver water according to a preset schedule programmed into your system. In cold regions yard irrigation systems get drained before winter, so a failure to operate could be caused by nothing more than a main water feed valve that needs to be turned ON. Home irrigation systems usually use proprietary hardware and fittings. If you need technical help, look for a brand name on your system, then call a service company that's familiar with your equipment.

THINGS YOU SHOULD KNOW ABOUT PLUMBING

Plumbing is often one of the most overlooked systems both in high-rise buildings and residential apartment blocks. Fixing a leaky faucet or a running toilet in your own house is one thing, but if you own a company that operates in the property management sector, plumbing issues can have grave effects both from a financial and business standpoint.

Whether you manage schools, office building, apartment blocks or healthcare facilities, you will not be spared by the occasional plumbing issue occurring and filling your inbox with complaints. And for good reason – faulty plumbing systems can affect the property in many ways, from permeating the building with unpleasant smells, helping mold develop in certain places by creating moisture, to degrading the overall structural integrity of the building.

Without further ado, here are ten things you should know about plumbing.

1. Automatic Leaks Detector

Taps, sinks, pipes, and other items may not last forever and may eventually weaken and start leaking. Some property managers let pipes stay for long after they have proven in disrepair, so eventually installing new ones at some point is essential. Assess the damages that the pipes and sinks have suffered and, if the situation calls of it, replace them with brand new ones. From a business standpoint, it is cheaper to replace old pipes than completely renovating the interior in the aftermath of a water-related accident. You can ask a professional plumber to inspect your water system and upgrade where necessary. Consider installing the latest systems which can detect leaks in property.

Moreover, keep in mind that the piping infrastructure in the United States is aging rapidly. In 2017, the average age of most pipes in the country was around 47 years old. The situation is worse in in New York and Philadelphia, where the pipes are pushing 70. Therefore, ensuring that the building has a working plumbing is vital.

2. Water Pressure

Water pressure is another plumbing aspect that experts should notice. Typically, in your average household, the water pressure should be under 80 psi.

The situation changes when it comes to bigger structures, such as high-rises or other public buildings. Buildings which are higher than eight stories require pumps, which transport the H2O into water tanks on the highest floor. This system is designed to ensure that water is distributed equally amongst all floors without sacrificing the level of water pressure.

On the other hand, high water pressure is one of the main causes of water-related accidents in all types of buildings, especially in smart homes. Moreover, since plumbing and construction technology has advanced, damages caused by flooding are even more expensive to repair.

3. Submeter

When it comes to large buildings like high-rises, submetering is the most practical approach. A submeter is an intricate system that allows landlords, condominium associations, landlords or other legal entities that manage buildings to charge tenants for individual consumption.

Submetering is the most practical approach regarding consumption measurement.

4. Prevention

As always, taking preventive measures is cheaper than handling the problems right after they occur. One way to prevent any plumbing accidents is to ensure that everything is correctly installed. Loose pipes or improper installation of plumbing traps for urinals are two of the most common causes of damage to the building. That is why regular retightening and check-ups are important to perform.

The same thing applies to toilets. If you are receiving complaints about broken toilets, it is cheaper on the long run to just replace them than doing shoddy patch-up works.

5. Plumbing Appliances

Each sink or toilet in any chosen building has a shut-off valve that allows the water supply to be cut off. Tour the property, locate the flaps and learn how to turn off the water supply in case toilets burst at night when there is no staff to attend to the situation.

Additionally, determine the location of the main water valve and how to operate it. In case you need to control or cut off the water supply, knowing how to operate it will prevent further damages from occurring.

6. Water System

Every building has turn of valves that are strategically spread around to provide easy access to them. By instructing and teaching your staff how to operate the water system, many future accidents can be prevented, which could be potentially costly from a business standpoint. Additionally, as a property manager, identify the limits concerning plumbing repairs your staff can handle and hire outside experts to do the job that you can't do.

7. Chemicals (Do Not Fix Water Problems)

Some property managers purchase chemicals and pour down drains to clean them. It's important to note that compounds might fix the problem, but at a cost. Chemicals weaken drain pipes and sinks and eventually cause damage. Sewerage lines and underground pipes often rot after exposure to chemicals.

Thankfully, experts have been working on new water cleaning technologies for years, so it is safe to assume this scenario might become obsolete sooner rather than later.

8. Consult a Professional Plumber

If the water system is experiencing persistent problems, reach out to a group of experts. The plumber might be expensive but will indeed save your business thousands of dollars in repairs. While training your staff to repair minor issues might be cheaper in the short term, outsourcing the job to external companies could prove to be more profitable on the long run.

9. Routine Maintenance Helps

Check your water lines on a regular basis. Keep an eye on the heater, wet walls, toilet, and the drops from the sink. Routine maintenance will reduce leakages and property damage from spreading further. Do not ignore any water issues, no matter how minor they might seem at a first glance. Small leakages can degenerate into bigger ones, requiring a full remake of the plumbing system.

10. Large Food Particles and Oils Block Pipes

Make sure to instruct any staff, office workers or tenants not to flush any grease into the kitchen sink. Bonney's CEO Allen Crick says that oils and grease can clog the pipes. The cooking oil should be poured into cartons and disposed into the trash. The same thing applies to food waste – potato peels can block the sink and lead to water spillage.

BEGINNER PLUMBING TIPS THAT EVERYONE SHOULD KNOW

Here are few tips and tricks that every do-it-yourselfer needs to know.

Our guys at Hillcrest Plumbing want you to know some things that they don't teach in high school but they can save you hours of time, hundreds of dollars, and certainly insurmountable amounts of frustration as you go about your plumbing projects as a beginner.

1.   PVC and CPVC pipes are two entirely different things. If you go to the store and do not know which one to ask for, you risk buying a pipe that will not be able to do the job that you need. Even the way we measure these pipes are different, so not only will the pipe likely not fit, they both do different things.

2.   CPVC pipe is much more preferable for hot water. Typically professional plumbers will use PVC pipe for cold water and regular water lines and CPVC pipe for hot water lines or things like a dishwasher or a washing machine that often have hot water running through them.

3.   They are measured differently. For PVC pipe, one would measure the diameter of the inside of the pipe. Whereas with CPVC pipe the measurement is taken based on the diameter of the outside of the pipe. A simple solution to make sure that you get exactly what you are looking for without having to go the store over and over again purchasing the wrong product, is to take a small piece of the

pipe that you are looking to fit and replace. The people at the hardware store will be able to tell you what type of pipe it is and the correct measurement.

4. Another fun fact: When dealing with PVC, you may find that there are two of everything! If you look a little bit closer at the fine print on the packaging there will be a little indicator to tell the two apart. One will have the indicator schedule 40 which is meant only for the use of pressurized hot water lines. The other indicator will read DWV. DVW indicates that these pipes are meant for drains, valves, and waterlines.

5. Though there are many different types of pipes, they do make universal cutters for all of types that may be worth investing in.

6. It doesn't stop with different types of piping. There are also different types of products for each kind of piping as well. So if you need glue, you need to make sure that the glue coordinates with the kind of pipe you intend on gluing, which of course means that you have to know that there are different

kinds of pipes and the kind of pipe that you are trying to glue.

7. Another thing to note before you glue. You have to first apply a primer. The primer is something that home inspectors check for to ensure that the job was done right. Primers will also have to correlate with the type of piping that you are priming.

8. Whatever you are working with whether it be the kitchen, the bathroom, or some other area of your home, you might consider adding a shut-off valve. Though it may seem like an unnecessary step, it certainly adds convenience in that you can shut off the water to a certain area of your home that you're working on without having to shut the water off to the entire house. So if you are changing the drain in the sink, you can still take a shower from the dirty job while the glue dries on your pipes in the kitchen.

9. Plumbing is very specific when it comes to home inspections. Local codes may have

variances, but inspectors do check to make sure that things are done right.

10.	The diameter of a shower drain must measure exactly 2" in diameter and maintain this diameter all the way to the main line in order for it to be in regulatory compliance.

11.	There is more flexibility for toilet drains. They can vary in diameter from 3"-4," but the larger one is always preferable and is less likely to have problems with clogs and back-ups.

12.	Vent pipes are required along the exterior of your home for all toilets. Each toilet must have one unless there are toilets that are close enough together that they are able to connect to the same vent pipe.

13.	Toilets must also each have a clean out, which allows for easy accessibility for a plumber to access and unclog back-ups and build-ups. The same rule applies here for multiple toilets. They must each have one, but can share a clean out if they are close enough to be connected to the same one.

14. Most sinks have what is called a P trap which is a removable pipe that is often the source of build-ups and clogs. Most drains can be unclogged by simply removing the P trap and taking out any debris and things that are caught inside.

15. Lower water pressure is often related to one of three things: a leaky faucet, clogged aerators, or a blockage in your pipes. All of which are very basic problems that can be tackled by most DIY's.

16. A water bill that jumps suddenly is often due to a leaky toilet or a toilet that is running constantly. While this isn't always the case, the majority of water expenses are due to the frequent use of water in bathrooms, thus indicating that if there is a significant increase there is a high likelihood that this is the source.

17. Constantly running toilets are toilets that seem to run all the time. This

issue is usually caused by a flapper that has gone bad. The flapper is a rubber part inside the back of the toilet that can often be found at the hardware store for under $10 and it only takes about ten minutes to replace.

18. Quite possibly the most common plumbing problem is that of a leaky faucet. These can also be caused by the wearing of an aerator, but could also indicate the need for replacement of a washer, rubber seal, or an o ring. These are all products that can be found at the hardware store for next to nothing and easily replaced by a beginning do-it-yourselfer with little hassle.

COMMON PLUMBING YOU SHOULD KNOW HOW TO DO AROUND THE HOUSE

Plumbing powers the essential utilities in our homes and enables us to accomplish daily and essential tasks, such as shower, drink water, cook, wash hands, brush teeth, flush the toilet, clean, heat water, treat air and more.

Most people don't give plumbing a second thought when it's working right, but it is all we can think about if something goes wrong. A basic understanding of your plumbing system and the components that affect it will help you troubleshoot, do small repairs yourself, know when to call a plumber, be better prepared in a crisis and make informed decisions.

A bit of general plumbing knowledge can save you money in service calls and prevent the headaches involved with breakdowns and problems.

1. Recognize the Source of Your Water

Generally, water comes into a home from one of two sources: a residential well and private pump or a city water line. Most of the time, rural residents have well water that is carried into the home via a pump, and they do not receive a water bill. Urban residents have city water they pay for by gallons of usage and usually receive a monthly or quarterly bill.

2. Test Water Quality

It is always good to know what's in the city or well water. Many people conduct tests when they move into a new place, but experts say to test well water at least once per year because much can change due to different supply or treatment, soil shifts and some processes used by agricultural or industrial businesses in the region.

You can purchase a water-testing kit from many types of suppliers, and many times the county entity in your area will offer them at a discount. Any drinking-water supplier is required to test the water annually and report on its quality, but people who have city water can conduct tests if they like.

You can shop around for the test you want from a trusted supplier, but you will find a variety available at costs that range from $45-$200. They reveal a number of different factors: nitrates, turbidity, heavy metals, bacteria, minerals, volatile organics and more. They come in consolidated kits that test for several common concerns at once. Most of the quality tests require you to capture some water and send it to a lab, which gives you results.

3. Locate and Turn Off Your Water Main

Should your home spring some kind of leak, you will appreciate knowing right where to go and what to do to cut off the water instead of trying to find it while you're panicked and water is spewing everywhere. There is almost always a main valve near the street, and sometimes a secondary in or around the house, such as in the basement. The water main usually resembles a wheel or bar-type lever. If it's a wheel, you should turn it slowly clockwise until it stops. If it's a lever, you push right (or down) until it stops.

If you're not sure the water is off, you can test it by trying to run water at a sink. If it does not run, you were successful. It never hurts to find the main and practice turning it off so that if a crisis occurs the process is familiar to you.

4. Find the Individual Cut-Off Valves

Check all your water-using appliances such as the washer, toilets and sinks to locate the small handle on them where you can stop their water supply. The washer handle is usually behind the item, near the wall. For the toilet, it's usually down low, toward the back and close to the wall. For sinks, look underneath near the pipes or against the walls. Showers and bathtubs are harder but might have cut-off valves in an adjacent closet or in the basement at the supply line. These individual valves enable you to isolate one place that may need maintenance without having to shut off water to the entire house.

5. Scrutinize the Water Meter and Bill

If you have city water, there is a meter for your individual home somewhere around it or perhaps near the street under a metal cover. Either way, knowing where your water meter is and how to read it will help you monitor usage and keep expenses down.

The water company can be a great help in locating the water meter and main shutoff, as well as to answer any questions you have about the bill. Many people want to know specifically how the water is metered, when the meters are checked and how much they're paying per gallon of water. Check the bill when it comes each month, because spikes in usage can indicate leaks or usage you don't see.

Look for 80 pounds per square inch of pressure as the household standard. You can ask your water company to test the pressure for you, or you can buy a water-pressure tester at most hardware or home-improvement stores starting for about $10. The gauge attaches to your outside water faucet and gives a reading of the water pressure when you turn on the spigot.

7. Adjust the Water Pressure

Anyone with municipal water has a water-pressure regulator between their home and the main supply. Without it, the pressure would blast through all of your appliances and fixtures. To raise or lower the pressure coming into your home, you need to find the regulator and have a partner to test the pressure as you adjust it.

The regulator might be in or around your home, near the main water-cutoff valve, or it could be near the street with the water main. It will probably have some kind of screwed-on cover, but inside is usually a wing nut or bolt you can turn to adjust the pressure. Again, while someone watches the pressure, slowly turn the nut-bolt clockwise to increase pressure and counterclockwise to decrease it.

8. Check for Hidden Leaks

Hidden leaks damage important infrastructure in your home like wooden beams, drywall, carpet and sometimes pipes and other plumbing accessories. You can test your system to see if it's using any excess water by checking your meter. Y don't have to wait until you think there is a problem. Regular checks can help you spot problems early.

First, make sure all indoor and outdoor spigots are turned off tightly. Pick a time when you will not need the water for at least 15 minutes or longer if possible — this includes any automatic appliances such as ice makers or water softeners and purifiers. Look at the numbers on your water meter and write them down. Then wait the time you've allotted and look at the numbers again when you return. If they are the same, your system is tight. If the numbers are higher, you likely have a leak you can't see. Sometimes the leak can be as simple as a toilet running that you haven't noticed yet, or as complicated as a small burst in a buried pipe. Sprinkler systems can often be the culprit of unknown or hidden leaks.

9. Get Acquainted With the Water Heater

Find the shutoff valves for water and for electric or gas supply on the water heater. For a gas water heater, turn the gas-supply line knob clockwise and for an electric water heater, find the breaker or fuse that supplies it and turn it off.

For the water supply, there's usually a handle or lever near where the water heater connects to the main water line. The water heater has two water lines. You want to turn off the cold, incoming water as opposed to the line that carries hot water out of the heater. Your water heater should have a temperature gauge and/or dial near the top or bottom, where you can adjust the temperature. The dial may have high, medium and low settings and others have a screw or other setup to adjust the temperature up or down. If you're unsure about adjusting the temperature, have a plumber do it for you.

Check all the pipes and accessories attached to the water heater regularly for leaks, and it never hurts to place some kind of drip pan underneath the hot-water heater. Water heaters typically last about 10-12 years, and they almost always start leaking toward the end of their life.

10. Change or Tighten a Toilet Seat

Behind the toilet seat between the tank and bowl are two bolts that hold the toilet seat in place. They usually have covers over them that match the color of the toilet. The rest of the fastening assembly is on the underneath side, but if your toilet seat is loose you can open the little doors over the bolts on the topside and turn them clockwise to see if that tightens it. If not, reach or look underneath and see if you feel or see anything loose.

You should be able to change the toilet seat by removing these two bolts, whether they fasten at the top, underneath or both. Though most holes are standard, be sure you pick a toilet seat that matches the bolt assembly in your current toilet.

11. Replace a Sink Stopper

bathroom sink stopper replacementIf the stopper in your bathroom sink doesn't work or breaks, you can easily replace the assembly or parts of it yourself. Clear out the underside of the sink, so you have room to work. Look for the rod-arm bar that holds the stopper in place and is normally attached to the sink drain pipe. Loosen the screw or whatever fastener is holding it in place, and you should be able to bring the sink stopper up for inspection and replacement if needed.

12. Swap Out a Shower Head

You can add customization in the bathroom that makes a big difference with a change in shower head. Some go for a more efficient model and others want a different kind of "stream," or maybe one that is removable for easy cleaning of the shower walls.

You can usually do the swap with a pair of common pliers, though in some cases a bigger size is needed. Before you put on the new shower head, be sure to wrap the threads with a few layers of thread-seal tape, which is usually thin, white and available where you buy the shower head. A similar process applies to change the shower arm, which holds the head to the wall and is usually sold separately.

13. Plunge a Toilet

Toilets can clog, but with the right kind of plunger or other tool, you need not worry. First, don't presume a tiny sink plunger will work in the toilet. You can buy a toilet plunger that is specially designed with a higher and more rounded hood and extended flange.

In short, a plunger uses manually produced suction to create pressure that clears the clog. You'll should make sure there is enough water in the bowl to cover the top of the plunger. If there's not, don't flush the toilet but take some water from the sink or bathtub to put on top of the plunger.

Put the plunger into the hole at the bottom of the toilet and thrust it up and down several times while maintaining its contact with the toilet bowl. This should cause the clog to clear, and you'll hear the toilet try to flush. If it doesn't, keep plunging until you do hear the toilet try to flush.

14. Replace a Toilet-Flapper Valve

If your toilet "runs," it could be because the flapper valve inside has become worn, chipped or damaged. If the seal isn't tight, water escapes the tank and it continuously tries to fill. You can usually remove the flapper valve with relative ease and take it to the home improvement store to find an exact-match replacement. Turn off the water supply to the toilet and then open the tank to find the flapper, which is usually a rubber piece that sits on top of a vertical column. If fixing a toilet doesn't sound appealing, it's always a good idea to consult a professional plumber!

15. Unclog a Sink

unclogging a sinkIf your sink drains slowly or is plugged and the traditional drain cleaner didn't do the trick, you can try a natural solution of 1/2 cup each of vinegar and baking soda. Mix it and pour it slowly into the drain to see if that helps move it.

Many times the drain is clogged at the U-shaped "p-trap" part of the plumbing pipe that's beneath the sink. First, turn off water to the sink and place a bucket underneath the p-trap before you unscrew both sides of the U to see what's inside and clear the clog. If you drop something down the sink like jewelry and need to retrieve it, don't run any water and follow this same procedure to retrieve the item out of the p-trap. If you're not sure you can handle this, it's always best to call the professionals than to risk further damage!

16. Locate Water and Sewer Lines

Whether you have a septic system and well or are connected to city water and sewer lines, it is useful to know where all of your utility lines run. Your city or water department should be able to provide a diagram of what it has on record or tell you their general location. If you have a septic system, you may have received an illustration of it and its lines at the time you bought the home.

17. Know Your Septic System

If you have a septic system, you should know what type you have and where all of its components are located. Some systems have easy-to-find parts while others may be buried, hidden or misidentified. You might have received a copy of a septic diagram at the time of home inspection, or there might be one on file at your county or other water-governing board in the area.

A professional plumber can help you positively identify the parts of your septic system, which can be helpful in the future.

Unless you are the person who installed the septic system, it is typically difficult to accurately locate all of its parts. You may have a mound system, drain field or holding-tank type of system that likely has access or clean-out points, vents and a straight-line connection to the home plumbing system. Usually a pump helps the system process as needed, but some older systems might work based on gravity alone.

18. Fix a Leaky Faucet

You might have a faucet that drips or leaks a bit. Sometimes the needed fix is a new o-ring or washer, which are the rubber seals found beneath the spigot base and each of the knobs. If you want to do it yourself, first turn the water off to the sink and stuff a rag into the drain, so no little parts fall down. Usually the screws to loosen the knobs are in them or underneath a cover you can pry off. The fasteners for the faucet are on the underside of the sink.

Once you loosen the assembly, you'll see the washers underneath the faucet handles and spigot. They can become dried or chipped and then leak water around the sink, or they can shut off loosely and leave a drip at the faucet. Take the old ones with you to the hardware store, so you can find an exact match for your sink.

19. Familiarize Yourself With the Breaker Panel

Anytime you decide on a do-it-yourself project that involves electricity, make sure you cut power to that part of the house before you start working. Somewhere in your home, there is an electrical panel of switches that controls the flow of electricity to different parts of the house. It usually has a metal door and is often in the basement, utility room or closet.

Most of the time, the parts of the home are labeled on the switches, but you can test and label the system yourself by the process of elimination. Turn each switch to its off position and see what part of your home does not have power. Repeat the process for each switch until they are labeled to your satisfaction. Some of them will be empty or inactive switches.

Some old homes have a fuse box that operates along the same principal as a breaker box. Different fuses help transfer power to different parts of the house and must be pulled out to cut power and replaced if they burn out.

20. Identify and Stop Drafts and Air Leaks

Any air that escapes your home or enters your home affects the efficiency of the heating, cooling and sometimes other systems. If you check for drafts and seal off any leaks you find, you can lower your bills. A stick of incense or a smoke pencil — which is available at the hardware store — can help you find leaks. Both emit a light smoke, and you carry it from room to room to hold it close to windows, doorways, structural joints, outlets, and other places such as underneath cabinets where air might enter or escape. If smoke escapes, use caulk, insulation or maybe replacement items to stop air flow.

21. Master the Caulk Gun

Caulk creates a seal that is needed around the house to lock out air or water. Some people prefer to invest in a caulk gun rather than running out of the smaller containers. With a caulk gun, it's easy to use and ready when you're sealing door thresholds, windows, around the bathroom tub or toilet base and other spots.

The gun will have a handle and a tip, and you'll need to snip the tip to a 45-degree angle to control the flow of caulk. Pull the gun handle back to install the tube. Soon you will have a feel for the steady-but-light pressure needed to create a clean and consistently sized line.

22. Protect Against a Frozen Spigot

Your plumber can help install a freeze-proof spigot, which basically extends to the inside of your home and connects to the water supply where it's warm instead of near the spigot where it freezes. The job involves pipe soldering, so it's usually best to have a professional do the work.

23. Venture into the Attic

You can catch many problems, both plumbing and roof related, early if you occasionally inspect your attic. Take a bright, powerful light and look up and around for any evidence of water, especially around the chimney, vents or any other place where there is an opening or things connect.

It is generally good to keep a few flashlights and fresh batteries for them. They're handy during a power outage, plus they're good for looking closely around attics, basements, crawl spaces and under sinks.

Chapter 15: Do It Yourself (Diy) Of Plumbing

You can become a do it yourself plumbing expert today and save hundreds of dollars, but you have to be careful. Minor plumbing problems come up almost on a daily basis. The biggest plumbing issues are most of the times result of carelessness. If you root out that leak or clog in time, you won't need a plumbing professional. However, some people simply get too excited about being their own plumbing superheroes, they mess up everything and end up with a plumber.

Today, we will give you a couple of valuable instructions that will prevent turning your do it yourself plumbing adventure into a disaster. When you know how to handle the materials properly and use the right tools, you will have so much fun fixing things around your house.

Do It Yourself Plumbing Instructions

In this part, we will talk about the initial steps towards a repair, then will mention the most common mistakes in do it yourself plumbing, and move on to a couple of incredibly interesting and easy do it yourself plumbing tricks.

Preparation

Locate the Valves

Always remember this when making any steps towards a do it yourself plumbing project – locate and shut off the valves. Take this seriously, even when performing the least important and least risky fixes (tightening toilet seat for example). If you have a nice and modern plumbing system in your home, there is a great possibility you have shutoff valves in each bathroom. When building a plumbing system, experts locate valves in the basement most of the times but don't rely on this only. It is always better to check if there are any hidden shutoffs behind access panels. The best test to check if you have turned off the right valves is to turn the faucet on or to flush the toilet. If there is any water running, then you should inspect the valves a bit better. The most reliable way to turn off the valves is to shut down the main one. You will find it easily, as the main valve is always located where the water line gets into your home. It may be that you will need to shut off the main valve someday in case of a water emergency, and this is a great opportunity to check on it in time.

Get the Sense of The Plumbing System in Your Home

If you are a complete novice in the field of plumbing, it would be great to do some research before hopping onto the work. When reading any of the step-by-step content, you should have at least a basic knowledge of what a vent pipe is, what is a hose clamp, or a sump pump. Even if you used to know a lot about plumbing and piping before, it would still be great to remind yourself of those concepts. Then, go around your house and inspect the main plumbing spots in it. You want to detect which walls in your home hide pipes or where the waste line is. Once you get familiar with the plumbing system in your home, it will be much easier to conduct a do it yourself plumbing project.

Prepare the Toolbox

As we mentioned above, you will need the right tools for managing any plumbing issue in your home. Of course, some repairs require additional purchases, but here are the tools you will need most of the times:

- Closet and/or hand auger,
- Adjustable wrench,
- Pipe wrench,
- Basin wrench,
- Hacksaw,
- Pliers,
- Metal file,
- Fire-resistant cloth

These tools should be near at hand for any plumbing repair, no matter if you want to do it yourself, or if you don't have another choice at a certain moment than to try to fix an issue yourself. After all, you don't want to get halfway through and then realize you are missing a tool in a starter kit.

Most Common Do It Yourself Mistakes

Dry Cleaner Overuse

When dealing with a major clog, chances are very low you are going to unclog it with a chemical drain cleaner. These products are effective with minor clogs, but you shouldn't overuse them. Any chemical causes a strong reaction in your pipes while spilling it all the time down the drain can only cause significant pipe damage instead of restoring the flow. Once you exceed the rational use of chemical drain cleaner, it will start building up somewhere down below. This will abrade the pipes bit by bit, but, what is even more important – can represent a threat to anyone who comes in contact with the chemicals. If, for any reason, built-up chemicals cause fumes in the pipes, that can cause breathing hazards. In case you are dealing with a major drain or you have already used a chemical drain cleaner recently, it would be much better to go for a baking soda, hot water, vinegar, and salt. Even if this doesn't work, you can try with an auger or a drain snake.

Overtightening

Pipes, fittings, toilet bolts, and supply tubes are very prone to cracking if you overtighten them. Most of the times, you won't notice the crank right away, but it will spread millimeter by millimeter, and cause a flood in a couple of weeks. Don't take things too rough, no matter if you are working with carbon or PVC pipes, as this can cost you a lot in the future.

Not Slanting the Shower Floor

If you, for any reason, wanted to replace the shower floor in your bathroom, remember that plumbing works according to the principle of gravity. One of the most common mistakes with do it yourself plumbing is not providing water to flow downward. People usually forget water flows only at a precise angle, and they create a perfect spot for molds and bacteria, intentionally. Take a slope while installing a shower floor. Ideally, you should slope it at a four percent angle.

Gluing

As I mentioned earlier, you want to be well informed when doing anything about plumbing. Remember that, when you want to seal something in your plumbing system, you don't need glue, but a plumber's putty. You wouldn't believe how many people make this mistake.

Forgetting about Measurements

Plumbing requires concentration and precision. If you made a cut on a pipe, you want to get a new piece that will fit into, right?

Another common mistake is not writing down measurements of the replacement part.

Tricks

Sweat The Pipes.

If you are a complete beginner with do it yourself plumbing, this trick is something you should remember for life. Once you learn how to sweat copper pipes, you passed the test.

First of all – you will like it because it is so easy. Secondly, you will learn how to fix smaller pipe issues in a very functional way. Listen, you won't make it look flawless, but you will get things done. If you are a real perfectionist, then you should better pay to an expert to fix the pipes.

However, you can't be completely confident about anyone actually. To perform this task, you would have to visit a local hardware shop and purchase a couple of items:

- Flux,

- Solder,

- A small torch (if you don't have one already),

- Pieces of copper pipe.

Now you have everything you need, and you are ready to unlock the carbon pipe mystery. Clean the pipes you want to fix and clean the copper pieces you bought. Next, heat the joint and take a little bit of flux. Once you do so, apply the solder. You see it is so simple, yet it will save your pockets for a long time. If you are trying this for the first time, you would probably be a bit clumsy, but don't worry – you will still fix the pipe. After all, the most important thing is that your joints would be nice and renovated.

Showerhead Issues

If you are dealing with a dripping showerhead, there is a great chance you can fix it within a couple of minutes. In most of the cases, dripping is a sign of improperly connected threads. Take the showerhead off, wrap tape in a clockwise direction, and take the head back.

On the other hand, if you noticed reduced water pressure coming out of your showerhead, that means only one thing – it is clogged. You will fix it easily if you take it off and soak in warm water and vinegar. Vinegar will abrade mineral deposits blocking the normal water spray.

Lose Pipe Fittings

A pipe wrench is not the answer in most of the cases. When it comes to plumbing, use your mind rather than your strength. Applying brute force won't lose hardened fittings. However, heat would, totally. Use a propane torch and apply the heat for a couple of minutes. Make sure there are no gas or plastic pipes in the area and cover your hands with a heat-resistant cloth. After you finish your work, make sure that you put isolation on your pipes to prevent freezing.

With these, you can handle at least 60% of plumbing repairs by yourself. Hence, there is no need for calling a professional for a small leak or a clog.

Common Problems in plumbing

Clogged lines

Clogs usually occur in drain lines, and need to be fixed promptly to avoid damage from a sink or toilet that overflows. Most toilet clogs can be dislodged using a plunger (aka plumber's helper). Sink clogs typically occur in the P-trap located directly below the sink drain. To clean out a clogged trap, place a bucket beneath the trap and detach the trap by unscrewing two slip nuts. Empty the trapped water in the bucket, clean out the trap, then reassemble your plumbing. Call in a plumber if you can't locate or dislodge a clog.

Leaks

Water leaks can cause major damage, but most leaks are preventable. Here are the major causes of leaks in a plumbing system:

Leaks around fittings

An improperly made solder joint can leak in copper plumbing lines. Compression or threaded plumbing connections can leak if the connection is loose. To prevent leaks in screwed connections (used on showerheads and flexible supply lines for sinks and toilets), wrap Teflon plumber's tape around the threads before attaching the connection nut.

Cracked pipe

PVC and CPVC pipe can crack from a hard impact or from water that freezes in the pipe. Copper pipe can crack if water freezes in a pipe run.

Faulty valves

Quality can vary in plumbing valves. A good valve can provide leak-free performance for many years, while a poorly made valve will start to leak after frequent use.

Impurities

Today, the water supply in many areas is more likely to contain impurities that can pose serious health hazards. The good news is that mitigation systems can remove this contamination if they're installed and maintained properly. Make sure to have your water tested by a water test company. These professionals can recommend the proper treatment to remove specific contaminants.

Excessive water use

Modern plumbing fixtures like toilets and shower-heads are designed to use water more efficiently than older fixtures. Replacing old, water-wasting fixtures with new, more efficient versions will lower your sewer bill (or reduce the load on your septic system), while also protecting the environment. Use the same strategy with your dishwasher and washing machine.

Well and septic problems

A house with its own well and septic system requires maintenance and occasional repairs. A septic tank should be pumped at least once a year, and care must be taken not to damage a septic field by running heavy equipment over the field area.

A well can sometimes be contaminated with bacteria. This will show up in a water test. If bacteria are present, the problem can be solved with a chlorine treatment. The well pump that delivers water to a holding tank will eventually require replacement, but this is a standard repair that any plumber can handle.

HOW TO CHOOSE A SUPPLIER FOR YOUR KITCHEN, BATHROOM AND PLUMBING SUPPLIES

If you are looking for plumbing supplies for your kitchen or bathroom, it is advisable you get them from a reputable supplier. A reputable supplier will sell you quality products that you will use for several years. However, most buyers find it a challenge choosing one supplier among the many, because the market is flooded with many of them. Here are factors you should consider when choosing a supplier for your kitchen and bathroom supplies.

Variety is the first factor you should consider when choosing a supplier. A reputable should have various items for your kitchen such as sink, taps, waste disposal units and many more. If it's the bathroom, they should have the showers and other bathroom accessories. It is important to also check if the supplier has the items in different designs, styles and colors.

Quality is another factor you should consider. If you want kitchen, bathroom and plumbing items that will last you for a long duration of time choose a supplier that has built a reputation over the years for selling quality products.

It is important to go online and check the reviews that have been done on the supplier you are planning to deal with. Read the comments that clients have left on different social media pages of the supplier. Check if there are any complains about the quality or customer service of the supplier.

Price is another factor you should also consider. The price of the kitchen, bathroom and plumbing products vary from one supplier to another, it is advisable to compare the prices from different suppliers. Go for a supplier that is selling the products at an affordable price without compromising on quality. If you are buying a variety of products, you can check if the supplier will sell for you the products at a wholesale price.

Another thing you should look out for when choosing a supplier, is the location. Go for a supplier based close to your area, it will be easier to transport the products to your place. In case you buy the supplies online, it is important to check the delivery period. Also, check the shipping cost, if you find a supplier who offers free shipping take advantage of that.

It is important to also check the manufacturer that supplies the kitchen, bathroom and plumbing supplies to the supplier. Go for a plumber who deals with reputable manufactures. Manufactures who are known to make quality and durable products.

BATHROOM REMODELING TIPS

Everyone wants a bathroom that shows what style and status they live. It has to be as comfortable as possible besides being appealing to the eye. Some people have dreams they have so far seen in the televisions and on magazines. All you have to know is that every idea is possible and achievable. Below are some of the tips that could prove to be useful when remodeling your bathroom.

The first thing is to come up with a bathroom planner. First things first, ensure that you talk to everyone that will be involved directly or indirectly in the bathroom space. Discuss the budget with them and make sure that you put aside some amount for emergency basing the fact that you never know what might happen shortly. You might want to change it or even it could have problems that you might want to fix later. Fix the time span that you want to use to do the remodeling process basing facts that time is money. Know that the more time you spend could increase the money you spend. Otherwise, make sure that all these plans are made with advice from a contractor.

From here it is essential to choose the layout you want. There are the standard ways that a bathroom is designed. It will include maybe two bathrooms one in the master bedroom and the other for the general use in the house. The shower will be fitted with a bathtub, a sink, and a toilet. If the bathroom is to be used by more than one person, then there is a great need to add another pan just for extra reasons that one could be in a hurry and needs to help themselves.

You can decide to fix a custom bathroom with several suite fixtures. With this, you are likely to have a lot of cabinets in the bathroom that can be used for the many useful reasons. If it is going to be your first time to do a bathroom remodeling, then there are some things that you might want to know. The first thing is that you need to keep your new plumbing close to where the old plumbing was if not replacing them. If the old fixtures and pipes are not well then, it is vital that you get what the professional thinks should be done. There is need to have overhead bathroom lights where there are a lot of ambient options that you can possibly go with. Other than that, it is essential to take into consideration the ventilation, fans, space among other essential factors that will make your bathroom remodeling worth the show.

Conclusion

Thank you again for purchasing this book!

I hope this book was able to help you to stop seeing plumbing chores as extremely difficult and start seeing them instead as easy-to-handle jobs.

you just learn and follow the instructions presented in this book.

The next step is to apply the tips you have read in this book on some common plumbing problems. Make sure to follow them to the letter as failure to do so can result in plumbing disasters that could have been prevented.

Studying all the information provided in this book will also help you save the money you could have otherwise spent on calling on the services of a plumber for jobs that you can easily do on your own. You will feel a sense of great fulfillment when your family and friends thank you for fixing their plumbing problems with your DIY knowledge and skills.

Thank you and good luck!

www.ingramcontent.com/pod-product-compliance
Lightning Source LLC
Chambersburg PA
CBHW050024130526
44590CB00042B/1874